T0260009

Multilevel Modeling Using Mplus

Aims and scope

Large and complex datasets are becoming prevalent in the social and behavioral sciences and statistical methods are crucial for the analysis and interpretation of such data. This series aims to capture new developments in statistical methodology with particular relevance to applications in the social and behavioral sciences. It seeks to promote appropriate use of statistical, econometric and psychometric methods in these applied sciences by publishing a broad range of reference works, textbooks and handbooks.

The scope of the series is wide, including applications of statistical methodology in sociology, psychology, economics, education, marketing research, political science, criminology, public policy, demography, survey methodology and official statistics. The titles included in the series are designed to appeal to applied statisticians, as well as students, researchers and practitioners from the above disciplines. The inclusion of real examples and case studies is therefore essential.

Published Titles

Analyzing Spatial Models of Choice and Judgment with R
David A. Armstrong II, Ryan Bakker, Royce Carroll, Christopher Hare, Keith T. Poole, and Howard Rosenthal

Analysis of Multivariate Social Science Data, Second Edition
David J. Bartholomew, Fiona Steele, Irini Moustaki, and Jane I. Galbraith

Latent Markov Models for Longitudinal Data
Francesco Bartolucci, Alessio Farcomeni, and Fulvia Pennoni

Statistical Test Theory for the Behavioral Sciences
Dato N. M. de Gruijter and Leo J. Th. van der Kamp

Multivariable Modeling and Multivariate Analysis for the Behavioral Sciences
Brian S. Everitt

Multilevel Modeling Using Mplus
W. Holmes Finch and Jocelyn E. Bolin

Multilevel Modeling Using R
W. Holmes Finch, Jocelyn E. Bolin, and Ken Kelley

Big Data and Social Science: A Practical Guide to Methods and Tools
Ian Foster, Rayid Ghani, Ron S. Jarmin, Frauke Kreuter, and Julia Lane

Ordered Regression Models: Parallel, Partial, and Non-Parallel Alternatives
Andrew S. Fullerton and Jun Xu

Bayesian Methods: A Social and Behavioral Sciences Approach, Third Edition
Jeff Gill

Multiple Correspondence Analysis and Related Methods
Michael Greenacre and Jorg Blasius

Applied Survey Data Analysis
Steven G. Heeringa, Brady T. West, and Patricia A. Berglund

Informative Hypotheses: Theory and Practice for Behavioral and Social Scientists
Herbert Hoijtink

Generalized Structured Component Analysis: A Component-Based Approach to Structural Equation Modeling
Heungsun Hwang and Yoshio Takane

Bayesian Psychometric Modeling
Roy Levy and Robert J. Mislevy

Statistical Studies of Income, Poverty and Inequality in Europe: Computing and Graphics in R Using EU-SILC
Nicholas T. Longford

Foundations of Factor Analysis, Second Edition
Stanley A. Mulaik

Linear Causal Modeling with Structural Equations
Stanley A. Mulaik

Age–Period–Cohort Models: Approaches and Analyses with Aggregate Data
Robert M. O'Brien

Handbook of International Large-Scale Assessment: Background, Technical Issues, and Methods of Data Analysis
Leslie Rutkowski, Matthias von Davier, and David Rutkowski

Generalized Linear Models for Categorical and Continuous Limited Dependent Variables
Michael Smithson and Edgar C. Merkle

Published Titles (continued)

Chapman & Hall/CRC
Statistics in the Social and Behavioral Sciences Series

Multilevel Modeling Using Mplus

W. Holmes Finch

Ball State University
Muncie, Indiana, USA

Jocelyn E. Bolin

Ball State University
Muncie, Indiana, USA

CRC Press
Taylor & Francis Group
Boca Raton   London   New York

CRC Press is an imprint of the
Taylor & Francis Group, an **informa** business
A CHAPMAN & HALL BOOK

Contents

Preface

The goal of this book is to provide you, the reader, with a comprehensive resource for the conduct of multilevel modeling using the Mplus software package. Multilevel modeling, sometimes referred to as hierarchical modeling, is a powerful tool that allows the researcher to account for data collected at multiple levels. For example, an educational researcher might gather test scores and measures of socioeconomic status (SES) for students who attend a number of different schools. The students would be considered level 1 sampling units, and the schools would be referred to as level 2 units. Ignoring the structure inherent in this type of data collection can, as we discuss in Chapter 2, lead to incorrect parameter and standard error estimates. In addition to modeling the data structure correctly, we will see in the following chapters that the use of multilevel models can also provide us with insights into the nature of relationships in our data that might otherwise not be detected. After reviewing standard linear models in Chapter 1, we will turn our attention to the basics of multilevel models in Chapter 2, before learning how to fit these models using Mplus in Chapters 3 through 5. Chapter 6 focuses on the use of multilevel modeling in the case of longitudinal data. Chapters 7 and 8 describe models for categorical dependent variables, first for single-level data, and then in the multilevel context. Similarly, Chapters 9 and 10 focus on latent variable models, first at one level, and then at multiple levels. Finally, we conclude in Chapter 11 with the Bayesian fitting of multilevel models. The datasets featured in this book are available at https://www.crcpress.com/Multilevel-Modeling-Using-Mplus/Finch-Bolin/p/book/9781498748247.

We hope that you find this book to be helpful as you work with multilevel data. Our goal is to provide you with a guidebook that will serve as the launching point for your own investigations in multilevel modeling. The Mplus code and discussion of its interpretation presented in this text should provide you with the tools necessary to gain insights into your own research, in whatever field it might be. We appreciate you for taking the time to read our work, and hope that you find it as enjoyable and informative to read as it was for us to write.

Authors

W. Holmes Finch is the George and Frances Ball distinguished professor of educational psychology, and professor in the Department of Educational Psychology at Ball State University, Muncie, Indiana, where he has been working since 2003. He earned his PhD at the University of South Carolina, Columbia, South Carolina, in 2002. Dr. Finch teaches courses in factor analysis, structural equation modeling, categorical data analysis, regression, multivariate statistics, and measurement to graduate students in psychology and education. His research interests are in the areas of multilevel models, latent variable modeling, methods of prediction and classification, and nonparametric multivariate statistics. He is also an Accredited Professional Statistician (PStat®).

Jocelyn E. Bolin is an associate professor in the Department of Educational Psychology at Ball State University, Muncie, Indiana, and earned her PhD in educational psychology at Indiana University Bloomington in 2009. Her dissertation consisted a comparison of statistical classification analyses under situations of training data misclassification. Dr. Bolin teaches courses on introductory and intermediate statistics, multiple regression analysis, and multilevel modeling for graduate students in social science disciplines. Her research interests include statistical methods for classification and clustering and use of multilevel modeling in the social sciences. She is a member of the American Psychological Association, the American Educational Research Association, and the American Statistical Association. Dr. Bolin is also an Accredited Professional Statistician (PStat®).

1

Linear Models

Statistical models provide powerful tools to researchers in a wide array of disciplines. Such models allow for the examination of relationships among multiple variables, which in turn can lead to a better understanding of the world. For example, sociologists use linear regression to gain insights into how factors such as ethnicity, gender, and level of education are related to an individual's income. Biologists can use the same type of model to understand the interplay between sunlight, rainfall, industrial runoff, and biodiversity in a rain forest. In addition, using linear regression, educational researchers can develop powerful tools for understanding the role that different instructional strategies have on student achievement. Apart from providing a path by which various phenomena can be better understood, statistical models can also be used as predictive tools. For example, econometricians might develop models to predict labor market participation given a set of economic inputs, whereas higher education administrators may use similar types of models to predict grade point average (GPA) for prospective incoming freshmen in order to identify those who might need academic assistance during their first year of college.

As can be seen from these few examples, statistical modeling is very important across a wide range of fields, providing researchers with tools for both explanation and prediction. Certainly, the most popular of such models over the past 100 years of statistical practice has been the general linear model (GLM). The GLM links a dependent, or outcome, variable to one or more independent variables and can take the form of such popular tools as analysis of variance (ANOVA) and regression. Given its popularity and utility, and the fact that it serves as the foundation for many other models, including the multilevel models featured in this book, we will start with a brief review of the linear model, particularly focusing on regression. This chapter will include a short technical discussion of linear regression models, followed by a description of how they can be estimated using Mplus. The technical aspects of this discussion are purposefully not highly detailed, as we focus on the model from a conceptual perspective. However, sufficient detail is presented so that the reader having only limited familiarity with the linear regression model will be provided with a basis for moving forward to multilevel models, and so that particular features of these more complex models that are shared with the linear model can be explicated. Readers particularly familiar with linear regression and with using R to conduct such analyses may elect to skip this chapter with no loss of understanding in future chapters.

Simple Linear Regression

As noted previously, the GLM framework serves as the basis for the multilevel models that we describe in subsequent chapters. Thus, in order to provide the foundation for the rest of the book, we will focus in this chapter on the linear regression model, although its form and function can easily be translated to ANOVA as well. The simple linear regression model in population form is

$$y_i = \beta_0 + \beta_1 x_i + \varepsilon_i \tag{1.1}$$

where:
  y_i is the dependent variable for individual i in the dataset
  x_i is the independent variable for subject i
  The terms β_0 and β_1 are the intercept and slope of the model, respectively

In a graphical sense, the intercept is the point where the line in Equation 1.1 crosses the y-axis at $x = 0$. It is also the mean, specifically the conditional mean, of y for individuals with a value of 0 on x, and it is this latter definition that will be most useful in actual practice. The slope, β_1, expresses the relationship between y and x. Positive slope values indicate that larger values of x are associated with correspondingly larger values of y, whereas negative slopes mean that larger x values are associated with smaller y's. Holding everything else constant, larger values of β_1 (positive or negative) indicate a stronger linear relationship between y and x. Finally, ε_i represents the random error inherent in any statistical model, including regression. It expresses the fact that for any individual, i, the model will not generally provide a perfect predicted value of y_i, denoted $\hat{y}_i$ and obtained by applying the regression model as

$$\hat{y}_i = \beta_0 + \beta_1 x_i \tag{1.2}$$

Conceptually, this random error is representative of all factors that might influence the dependent variable other than x.

Estimating Regression Models with Ordinary Least Squares

In virtually all real-world contexts, the population is unavailable to the researcher. Therefore, β_0 and β_1 must be estimated using sample data taken from the population. There exist in the statistical literature several methods for obtaining estimated values of the regression model parameters (b_0 and b_1, respectively) given a set of x and y. By far, the most popular and widely used of these methods is ordinary least squares (OLS). The vast majority of other approaches are useful in special cases involving small samples or data that do not conform to the distributional assumptions undergirding OLS. The goal of OLS is to minimize the sum of the squared differences between the observed values of y and the model predicted values of y across the sample. This difference, known as the residual, is written as

$$e_i = y_i - \hat{y}_i \tag{1.3}$$

Therefore, the method of OLS seeks to minimize

$$\sum_{i=1}^{n} e_i^2 = \sum_{i=1}^{n} (y_i - \hat{y}_i)^2 \tag{1.4}$$

The actual mechanism for finding the linear equation that minimizes the sum of squared residuals involves the partial derivatives of the sum of squared function with respect to the model coefficients, β_0 and β_1. We will leave these mathematical details to excellent references, such as Fox (2008). It should be noted that in the context of simple linear regression, the OLS criteria reduces to the following equations, which can be used to obtain b_0 and b_1 as

$$b_1 = r \left(\frac{s_y}{s_x} \right) \tag{1.5}$$

and

$$b_0 = \bar{y} - b_1 \bar{x} \tag{1.6}$$

where:
  r is the Pearson product moment correlation coefficient between x and y
  s_y is the sample standard deviation of y
  s_x is the sample standard deviation of x
  $\bar{y}$ is the sample mean of y
  $\bar{x}$ is the sample mean of x

Distributional Assumptions Underlying Regression

The linear regression model rests upon several assumptions about the distribution of the residuals in the broader population. Although the researcher can typically never collect data from the entire population, it is possible to assess empirically whether these assumptions are likely to hold true based on the sample data. The first assumption that must hold true for linear models to function optimally is that the relationship between y_i and x_i is linear. If the relationship is not linear, then clearly an equation for a line will not provide adequate fit and the model is thus misspecified. A second assumption is that the variance in the residuals is constant regardless of the value of x_i. This assumption is typically referred to as homoscedasticity and is a generalization of the homogeneity of error variance assumption in ANOVA. Homoscedasticity implies that the variance of y_i is constant across the values of x_i. The distribution of the dependent variable around the regression line is literally the distribution of the residuals, thus making clear the connection of homoscedasticity of errors with the distribution of y_i around the regression line. The third such assumption is that the residuals are normally distributed in the population. Fourth, it is assumed that the independent variable x is measured without error and that it is unrelated to the model error term, ε. It should be noted that the assumption of x measured without error is not as strenuous as one might first assume. In fact, for most real-world problems, the model will work well even when the independent variable is not error free (Fox, 2008). Fifth and finally, the residuals for any two individuals in the population are assumed to be independent of one another. This independence assumption implies that the unmeasured factors influencing y are not related from one individual to another. It is this assumption that is directly addressed with the use of multilevel models, as we will see in Chapter 2. In many research situations, individuals are sampled in clusters, such that we cannot assume that individuals from the same such cluster will have uncorrelated residuals. For example, if samples are obtained from multiple neighborhoods, individuals within the same neighborhoods may tend to be more like one another than they are like individuals from other neighborhoods. A prototypical example of this is children within schools. Due to a variety of factors, children attending the same school will often have more in common with one another than they do with children from other schools. These "common" things might include neighborhood socioeconomic status, school administration policies, and school learning environment, to name just a few. Ignoring this clustering, or not even realizing it is a problem, can be detrimental to the results of statistical modeling. We explore this issue in great detail later in the book, but for now we simply want to mention that a failure to satisfy the assumption of independent errors is (1) a major problem but (2) often something that can be overcome with the appropriate models, such a multilevel models that explicitly consider the nesting of the data.

Coefficient of Determination

When the linear regression model has been estimated, researchers generally want to measure the relative magnitude of the relationship between the variables. One useful tool for ascertaining the strength of relationship between

x and y is the coefficient of determination, which is the squared multiple correlation coefficient, denoted R^2 in the sample. R^2 reflects the proportion of the variation in the dependent variable that is explained by the independent variable. Mathematically, R^2 is calculated as

$$R^2 = \frac{SS_R}{SS_T} = \frac{\sum\limits_{i=1}^{n}(\hat{y}_i - \bar{y})^2}{\sum\limits_{i=1}^{n}(y_i - \bar{y})^2} = 1 - \frac{\sum\limits_{i=1}^{n}(y_i - \hat{y})^2}{\sum\limits_{i=1}^{n}(y_i - \bar{y})^2} = 1 - \frac{SS_E}{SS_T} \qquad (1.7)$$

The terms in the above equation are as defined previously. The value of this statistic always lies between 0 and 1, with larger numbers indicating a stronger linear relationship between x and y, implying that the independent variable is able to account for more variance in the dependent. R^2 is a very commonly used measure of the overall fit of the regression model and, along with the parameter inference discussed next, serves as the primary mechanism by which the relationship between the two variables is quantified.

Inference for Regression Parameters

A second method for understanding the nature of the relationship between x and y involves making inferences about the relationship in the population given the sample regression equation. Because b_0 and b_1 are sample estimates of the population parameters β_0 and β_1, respectively, they are subject to sampling error as is any sample estimate. This means that, although the estimates are unbiased given that the aforementioned assumptions hold, they are not precisely equal to the population parameter values. Furthermore, were we to draw multiple samples from the population and estimate the intercept and slope for each, the values of b_0 and b_1 would differ across samples, even though they would be estimating the same population parameter values for β_0 and β_1. The magnitude of this variation in parameter estimates across samples can be estimated from our single sample using a statistic known as the standard error. The standard error of the slope, denoted as σ_{b_1} in the population, can be thought of as the standard deviation of slope values obtained from all possible samples of size n, taken from the population. Similarly, the standard error of the intercept, σ_{b_0}, is the standard deviation of the intercept values obtained from all such samples. Clearly, it is not possible to obtain census data from a population in an applied research context. Therefore, we will need to estimate the standard errors of both the slope (s_{b_1}) and intercept (s_{b_0}) using data from a single sample, much as we did with b_0 and b_1.

In order to obtain s_{b_1}, we must first calculate the variance of the residuals:

$$s_e^2 = \frac{\sum\limits_{i=1}^{n} e_i^2}{n-p-1} \tag{1.8}$$

where:

e_i is the residual value for individual i

N is the sample size

p is the number of independent variables (1 in the case of simple regression)

Then

$$s_{b_1} = \frac{1}{\sqrt{1-R^2}} \left[\frac{s_e}{\sqrt{\sum\limits_{i=1}^{n}(x_i - \bar{x})^2}} \right] \tag{1.9}$$

The standard error of the intercept is calculated as

$$s_{b_0} = s_{b_1} \sqrt{\frac{\sum\limits_{i=1}^{n} x_i^2}{n}} \tag{1.10}$$

Given that the sample intercept and slope are only estimates of the population parameters, researchers are quite often interested in testing hypotheses to infer whether the data represent a departure from what would be expected in what is commonly referred to as the null case, that the idea of the null value holding true in the population can be rejected. Most frequently (though not always), the inference of interest concerns testing that the population parameter is 0. In particular, a non-0 slope in the population means that x is linearly related to y. Therefore, researchers typically are interested in using the sample to make inference about whether the population slope is 0 or not. Inference can also be made regarding the intercept, and again the typical focus is on whether this value is 0 in the population.

Inference about regression parameters can be made using confidence intervals and hypothesis tests. Much as with the confidence interval of the mean, the confidence interval of the regression coefficient yields a range of values within which we have some level of confidence (e.g., 95%) that the population parameter value resides. If our particular interest is in whether x is linearly related to y, then we would simply determine whether 0 is in the interval for β_1. If so, then we would not be able to conclude that the population value differs from 0. The absence of a statistically significant result (i.e., an interval not containing 0) does not imply that the null hypothesis is true, but rather it means that there is not sufficient evidence available in the sample data to reject the null. Similarly, we can construct a confidence interval for the intercept, and if 0 is within the interval, we would conclude that the value of y for an individual with $x = 0$ could plausibly be, but is not necessarily 0. The confidence intervals for the slope and intercept take the following forms:

$$b_1 \pm t_{cv}s_{b_1} \tag{1.11}$$

and

$$b_0 \pm t_{cv}s_{b_0} \tag{1.12}$$

Here the parameter estimates and their standard errors are as described previously, whereas t_{cv} is the critical value of the t distribution for $1 - \alpha/2$ (e.g., the 0.975 quantile if $\alpha = 0.05$) with $(n - p - 1)$ degrees of freedom. The value of α is equal to 1 minus the desired level of confidence. Thus, for a 95% confidence interval (0.95 level of confidence), α would be 0.05.

In addition to confidence intervals, inference about the regression parameters can also be made using hypothesis tests. In general, the forms of this test for the slope and intercept respectively are

$$t_{b_1} = \frac{b_1 - \beta_1}{s_{b_1}} \tag{1.13}$$

$$t_{b_0} = \frac{b_0 - \beta_0}{s_{b_0}} \tag{1.14}$$

The terms β_1 and β_0 are the parameter values under the null hypothesis. Again, most often the null hypothesis posits that there is no linear relationship between x and y ($\beta_1 = 0$) and that the value of $y = 0$ when $x = 0$ ($\beta_0 = 0$). For simple regression, each of these tests is conducted with $(n - 2)$ degrees of freedom.

Multiple Regression

The linear regression model can very easily be extended to allow for multiple independent variables at once. In the case of two regressors, the model takes the form:

$$y_i = \beta_0 + \beta_1 x_{1i} + \beta_2 x_{2i} + \varepsilon_i \tag{1.15}$$

In many ways, this model is interpreted as is that for simple linear regression. The only major difference between simple and multiple regression interpretation is that that each coefficient is interpreted in turn *holding constant* the value of the other regression coefficient. In particular, the parameters are estimated by b_0, b_1, and b_2, and inferences about these parameters are made in the same fashion with regard to both confidence intervals and hypothesis tests. The assumptions underlying this model are also the same as those described for the simple regression model. Despite these similarities, there are three additional topics regarding multiple regression that we need to consider here. These are inferences for the set of model slopes as a whole, an adjusted measure of the coefficient of determination, and the issue of collinearity among the independent variables. Because these issues will be important in the context of multilevel modeling as well, we will address them in detail here.

With respect to model inference, for simple linear regression the most important parameter is generally the slope, so that inference for it will be of primary concern. When there are multiple x variables in the model, the researcher may want to know whether the independent variables taken as a whole are related to y. Therefore, some overall test of model significance is desirable. The null hypothesis for this test is that all of the slopes are equal to 0 in the population; that is, none of the regressors are linearly related to the dependent variable. The test statistic for this hypothesis is calculated as

$$F = \frac{SS_R/p}{SS_E/(n-p-1)} = \left(\frac{n-p-1}{p}\right)\left(\frac{R^2}{1-R^2}\right) \qquad (1.16)$$

Here terms are as defined in Equation 1.7. This test statistic is distributed as an F with p and $(n - p - 1)$ degrees of freedom. A statistically significant result would indicate that one or more of the regression coefficients are not equal to 0 in the population. Typically, the researcher would then refer to the tests of individual regression parameters, which were described previously, in order to identify which were not equal to 0.

A second issue to be considered by researchers in the context of multiple regression is the notion of adjusted R^2. Stated simply, the inclusion of additional independent variables in the regression model will always yield higher values of R^2, even when these variables are not statistically significantly related to the dependent variable. In other words, there is a capitalization on chance that occurs in the calculation of R^2. As a consequence, models including many regressors with negligible relationships with the y may produce an R^2 that would suggest the model explains a great deal of variance in y. An option for measuring the variance explained in the dependent variable that accounts for this additional model complexity would be quite helpful to the researcher seeking to understand the true nature of the relationship between the set of independent variables and the dependent. Such a measure exists in the form of the adjusted R^2 value, which is commonly calculated as

$$R_A^2 = 1 - \left(1 - R^2\right)\left(\frac{n-1}{n-p-1}\right) \qquad (1.17)$$

R_A^2 only increases with the addition of an x if that x explains more variance than would be expected by chance. R_A^2 will always be less than or equal to the standard R^2. It is generally recommended to use this statistic in practice when models containing many independent variables are used.

A final important issue specific to multiple regression is that of collinearity, which occurs when one independent variable is a linear combination of one or more of the other independent variables. In such a case, regression coefficients and their corresponding standard errors can be quite unstable, resulting in poor inference. It is possible to investigate the presence of collinearity using a statistic known as the variance inflation factor (VIF). In order to obtain the VIF for x_j, we would first regress all of the other independent variables onto x_j and obtain an $R_{x_i}^2$ value. We then calculate

$$\text{VIF} = \frac{1}{1 - R_x^2} \tag{1.18}$$

The VIF will become large when $R_{x_j}^2$ is near 1, indicating that x_j has very little unique variation when the other independent variables in the model are considered. That is, if the other $(p - 1)$ regressors can explain a high proportion of x_j, then x_j does not add much to the model, above and beyond the other $(p - 1)$ regression. Collinearity in turn leads to high sampling variation in b_j, resulting in large standard errors and unstable parameter estimates. Conventional rules of thumb have been proposed for determining when an independent variable is highly collinear with the set of other $(p - 1)$ regressors. Thus, the researcher might consider collinearity to be a problem if VIF > 5 or 10 (Fox, 2008). The typical response to collinearity is to either remove the offending variable(s) or use an alternative approach to conducting the regression analysis such as ridge regression or regression following a principal components analysis.

Example of Simple Linear Regression by Hand

In order to demonstrate the principles of linear regression discussed previously, let us consider a simple scenario in which a researcher has collected data on college GPA and test anxiety using a standard measure where higher scores indicate greater anxiety when taking a test. The sample consisted of 440 college students who were measured on both variables. In this case, the researcher is interested in the extent to which test anxiety is related to college GPA, so that GPA is the dependent and anxiety is the independent variable. The descriptive statistics for each variable, as well as the correlation between the two appear in Table 1.1.

TABLE 1.1

Descriptive Statistics and Correlation for GPA and Test Anxiety

Variable	Mean	Standard Deviation	Correlation
GPA	3.12	0.51	−0.30
Anxiety	35.14	10.83	

We can use this information to obtain estimates for both the slope and the intercept of the regression model using Equations 1.4 and 1.5. First, the slope is calculated as $b_1 = -0.30(0.51/10.83) = -0.014$, indicating that individuals with higher test anxiety scores will generally have lower GPAs. Next, we can use this value and information in the table to calculate the intercept estimate: $b_0 = 3.12 - (-0.014)(35.14) = 3.63$.

The resulting estimated regression equation is then $\hat{GPA} = 3.63 - 0.014$. (anxiety). Thus, this model would predict that for a 1-point increase in the anxiety assessment score, GPA would decrease by −0.014 points.

In order to better understand the strength of the relationship between test anxiety and GPA, we will want to calculate the coefficient of determination. To do this, we need both the SS_R and SS_T, which take the values 10.65 and 115.36, respectively, yielding $R^2 = 10.65/115.36 = 0.09$. This result suggests that approximately 9% of the variation in GPA is explained by variation in test anxiety scores. Using this R^2 value and Equation 1.14, we can calculate the F statistic testing whether any of the model slopes (in this case there is only one) are different from 0 in the population: $F = [(440-1-1)/1](0.09/(1-0.09)) = 438(0.10) = 43.80$. This test has p and $(n - p - 1)$ degrees of freedom, or 1 and 438 in this situation. The p-value of this test is less than .001, leading us to conclude that the slope in the population is indeed significantly different from 0 because the p-value is less than the Type I error rate specified. Thus, test anxiety is linearly related to GPA. The same inference could be conducted using the t-test for the slope. First we must calculate the standard error of the slope estimate: $s_{b_1} = [1/(\sqrt{1-R^2})]\{S_E/[\sqrt{\sum(x_i - \bar{x})^2}]\}$.

For these data $S_E = \sqrt{104.71/(440-1-1)} = \sqrt{0.24} = 0.49$. In turn, the sum of squared deviations for x (anxiety) was 53743.64, and we previously calculated $R^2 = 0.09$. Thus, the standard error for the slope is $s_{b_1} = (1/\sqrt{1-0.09}) \times (0.49/\sqrt{53743.64}) = 1.05(0.002) = 0.002$. The test statistic for the null hypothesis that $\beta_1 = 0$ is calculated as $t = (b_1 - 0)/s_{b_1} = -0.014/0.002 = -7.00$, with $(n - p - 1)$ or 438 degrees of freedom. The p-value for this test statistic value is less than .001, and thus, we can probabilistically infer that in the population, the value of the slope is not zero, with the best sample point estimate being −0.014.

Finally, we can also draw inference about β_1 through a 95% confidence interval, as shown in Equation 1.9. For this calculation, we will need to determine the value of the t distribution with 438 degrees of freedom that corresponds to the $1 - 0.05/2$ or 0.975 point in the distribution. We can do so by using a t table in the back of a textbook, or through standard computer software such as SPSS (IBM, 2013). In either case, we find that the critical value for this example is 1.97. The confidence interval can then be calculated as

$$\left(-0.014 - 1.97\left(0.002\right), -0.014 + 1.97\left(0.002\right)\right)$$

$$\left(-0.014 - 0.004, -0.104 + 0.004\right)$$

$$\left(-0.018, -0.010\right)$$

The fact that 0 is not in the 95% confidence interval simply supports the conclusion we reached using the p-value as described previously. Also, given this interval, we can infer that the actual population slope value lies between -0.018 and -0.010. Thus, anxiety could plausibly have an effect as small as little as -0.01 or as large as -0.018.

Regression Using Mplus

Returning to the previous example, predicting GPA from measures of physical (BStotal) and cognitive academic anxiety (CTA.tot), the program for fitting a multiple regression model in Mplus appears as follows:

```
TITLE:     Model 1.1 Multiple regression model
DATA: FILE IS achieve.csv;
VARIABLE: NAMES ARE Male Minority GPA BStotal CTA_tot;
           usevariables are GPA BStotal CTA_tot;
           missing are .;
MODEL:     GPA ON BStotal CTA_tot;
OUTPUT: standardized;
```

It is worth taking a little time to describe how this program is constructed. The first line consists of a title that we can use to explain the purpose of the program and any other information that might be helpful to the programmer. Next, the file is read in on the DATA line. The name of the file is achieve.csv and is saved in the same directory as the program file itself; thus, no further directory information is required. The data file itself is comma delimited (variables are separated by commas), which Mplus is able to read automatically. The variables contained in the data set are next listed, as are the variables that we will actually use in the analysis, GPA, BStotal, and CTA_tot. In addition, we define the value used for missing data, a . in this case. On the MODEL line, we indicate the dependent variable (GPA) and the independent variables (BStotal and CTA_tot). Finally, on the OUTPUT line, we request

standardized parameter estimates, which will provide us with beta weights.
The output obtained from this program appears as follows:

```
MODEL RESULTS
```

	Estimate	S.E.	Est./S.E.	Two-Tailed P-Value
GPA ON				
BSTOTAL	0.013	0.005	2.662	0.008
CTA_TOT	-0.020	0.003	-6.570	0.000
Intercepts				
GPA	3.619	0.079	45.793	0.000
Residual Variances				
GPA	0.234	0.016	14.646	0.000

```
STANDARDIZED MODEL RESULTS

STDYX Standardization
```

	Estimate	S.E.	Est./S.E.	Two-Tailed P-Value
GPA ON				
BSTOTAL	0.172	0.064	2.680	0.007
CTA_TOT	-0.425	0.062	-6.882	0.000
Intercepts				
GPA	7.074	0.233	30.397	0.000
Residual Variances				
GPA	0.893	0.028	31.713	0.000
Residual Variances				
GPA	0.893	0.028	31.575	0.000

```
R-SQUARE
```

Observed Variable	Estimate	S.E.	Est./S.E.	Two-Tailed P-Value
GPA	0.107	0.028	3.792	0.000

Based on these results, we can see that there is a statistically significant positive relationship between scores on physical anxiety (BSTOTAL) and GPA, and a negative relationship between scores on cognitive anxiety scores (CTA_tot) and GPA. The standardized coefficients for BSTOTAL and CTA_tot are 0.172 and -0.426, respectively. The R^2 for the model is 0.107, meaning that 10.7% of the variance in GPA is associated with physical and cognitive anxiety scores.

Interaction Terms in Regression

More complicated regression relationships can also be easily modeled using Mplus. Let us consider a moderation analysis involving the anxiety measures. In this example, an interaction between cognitive test anxiety and physical anxiety is modeled in addition to the main effects for the two variables. An interaction is simply computed as the product of the interacting variables, using the define command, as in the program that follows. In all other respects, the program for Model 1.2 is identical to that for Model 1.1.

```
TITLE:    Model 1.2 Multiple regression model with interaction
DATA: FILE IS achieve.csv;
VARIABLE: NAMES ARE Male Minority GPA BStotal CTA_tot;
          usevariables are GPA BStotal CTA_tot interaction;
          missing are .;
define:
          interaction = BStotal*CTA_tot;
MODEL:    GPA ON BStotal CTA_tot interaction;
OUTPUT: standardized;
```

MODEL RESULTS

	Estimate	S.E.	Est./S.E.	Two-Tailed P-Value
GPA ON				
BSTOTAL	-0.006	0.016	-0.367	0.714
CTA_TOT	-0.027	0.006	-4.444	0.000
INTERACTION	0.000	0.000	1.293	0.196
Intercepts				
GPA	3.898	0.230	16.971	0.000
Residual Variances				
GPA	0.233	0.016	14.646	0.000

STANDARDIZED MODEL RESULTS

STDYX Standardization

	Estimate	S.E.	Est./S.E.	Two-Tailed P-Value
GPA ON				
BSTOTAL	-0.074	0.201	-0.367	0.714
CTA_TOT	-0.568	0.125	-4.535	0.000
INTERACTION	0.366	0.283	1.295	0.195
Intercepts				
GPA	7.619	0.471	16.174	0.000

```
Residual Variances
    GPA                    0.890      0.029      31.210         0.000

R-SQUARE

    Observed                                              Two-Tailed
    Variable           Estimate      S.E.    Est./S.E.     P-Value

    GPA                    0.110      0.029       3.862        0.000
```

Here, the slope for the interaction is denoted interaction, takes the value 0.000, and is nonsignificant ($t = 1.293$, $p = .196$), indicating that the level of physical anxiety symptoms (BStotal) does not change or moderate the relationship between cognitive test anxiety (CTA.tot) and GPA.

Categorical Independent Variables

Mplus is also easily capable of incorporating categorical variables into analyses, including regression. Let us consider an analysis where we predict GPA from cognitive test anxiety (CTA.tot) and the categorical variable, gender. In order to incorporate gender into the model, it must be dummy coded such that one category (e.g., male) takes the value of 1, and the other category (e.g., female) takes the value of 0. In this example, we have named the variable Male, where 1 = male and 0 = not male (female). Defining a model using a dummy variable in Mplus then becomes no different from using continuous predictor variables.

```
TITLE:     Model 1.3 Multiple regression model with a
categorical independent variable
DATA:    FILE IS achieve.csv;
VARIABLE: NAMES ARE Male Minority GPA BStotal CTA_tot;
          usevariables are GPA CTA_tot Male;
          missing are .;
MODEL:     GPA ON CTA_tot Male;
OUTPUT: standardized; summary (Model1.3)

MODEL RESULTS
                                                      Two-Tailed
                   Estimate      S.E.    Est./S.E.     P-Value

GPA        ON
    CTA_TOT          -0.015      0.002      -7.197        0.000
    MALE             -0.223      0.047      -4.737        0.000
```

```
        Intercepts
          GPA                3.740       0.081       46.369        0.000

        Residual Variances
          GPA                0.226       0.015       14.832        0.000
```

STANDARDIZED MODEL RESULTS

STDYX Standardization

```
                                                          Two-Tailed
                         Estimate      S.E.     Est./S.E.   P-Value

        GPA         ON
          CTA_TOT          -0.320       0.042      -7.572       0.000
          MALE             -0.210       0.044      -4.832       0.000

        Intercepts
          GPA               7.305       0.226      32.369       0.000

        Residual Variances
          GPA               0.864       0.030      28.398       0.000
```

R-SQUARE

```
          Observed                   Two-Tailed
          Variable       Estimate      S.E.     Est./S.E.   P-Value

          GPA             0.136        0.030       4.485       0.000
```

In this example, the slope for the dummy variable Male is negative and significant ($\beta = -0.223$, $p < .001$), indicating that males have a significantly lower mean GPA than do females.

Using the define command, we can create an interaction term involving both categorical and continuous variables. In the following example, we will fit a model with an interaction between the variables Male and CTA_tot. The interaction term can be used to test whether the relationship of CTA_tot to GPA is the same for males and females.

```
TITLE:     Model 1.4 Multiple regression model with a
categorical independent variable
    and an interaction
DATA: FILE IS achieve.csv;
VARIABLE:  NAMES ARE Male Minority GPA BStotal CTA_tot;
           usevariables are GPA CTA_tot Male interaction;
           missing are .;
define:
           interaction = Male*CTA_tot;
MODEL:     GPA ON CTA_tot Male interaction;
OUTPUT: standardized;
```

MODEL RESULTS

	Estimate	S.E.	Est./S.E.	Two-Tailed P-Value
GPA ON				
CTA_TOT	-0.016	0.003	-6.076	0.000
MALE	-0.295	0.160	-1.846	0.065
INTERACTION	0.002	0.004	0.476	0.634
Intercepts				
GPA	3.767	0.098	38.460	0.000
Residual Variances				
GPA	0.226	0.015	14.832	0.000

STANDARDIZED MODEL RESULTS

STDYX Standardization

	Estimate	S.E.	Est./S.E.	Two-Tailed P-Value
GPA ON				
CTA_TOT	-0.335	0.053	-6.297	0.000
MALE	-0.279	0.151	-1.851	0.064
INTERACTION	0.072	0.152	0.476	0.634
Intercepts				
GPA	7.356	0.249	29.582	0.000
Residual Variances				
GPA	0.863	0.030	28.352	0.000

R-SQUARE

Observed Variable	Estimate	S.E.	Est./S.E.	Two-Tailed P-Value
GPA	0.137	0.030	4.495	0.000

There is not a statistically significant interaction between gender and cognitive anxiety, indicating that the relationship between anxiety and GPA is the same for males and females. Thus, the optimal model is one that does not include the interaction of the variables with one another.

Checking Regression Assumptions with Mplus

When checking assumptions for linear regression models, it is often desirable to create a plot of the residuals. Diagnostic residuals plots can be easily obtained through the use of the plot line in our Mplus program. Let us return to Model 1.1 predicting GPA from cognitive test anxiety and physical anxiety symptoms. By including the following command at the bottom of the program, we obtain the ability to carry out plots in Mplus, using the menu:

```
plot:    type=plot3;
```

After running this program, we can now access the plots through the menu bar in the icon ☑. By clicking on this button, we obtain the following window.

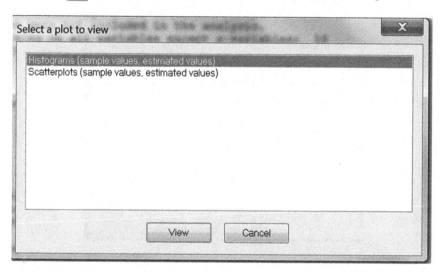

In order to obtain the scatterplot of observed and predicted values, we would double click on Scatterplots (sample values, estimated values).

In order to assess the assumption of variance homogeneity, we will want to examine a plot with the model-predicted GPA values on the *x*-axis and the residuals on the *y*-axis (Figure 1.1).

The assumption of homogeneity of variance can be checked through an examination of the residual by fitted plot. If the assumption holds, this plot should display a formless cloud of data points with no discernible shape and that are equally spaced across all values of *x*.

The linearity of the relationships between each independent variable and the dependent is assessed through an examination of the plots involving them. We can obtain such plots using the sequence of menu selections described previously, and placing each of the independent variables in turn on the *x*-axis and the residuals on the *y*-axis. Figure 1.2 shows plots for both BStotal and CTA_tot.

It is appropriate to assume linearity for an independent variable if the residual plots show no discernible pattern, which appears to be the case for both BStotal and CTA_tot.

In addition to linearity and homogeneity of variance, it is also important to determine whether the residuals follow a normal distribution,

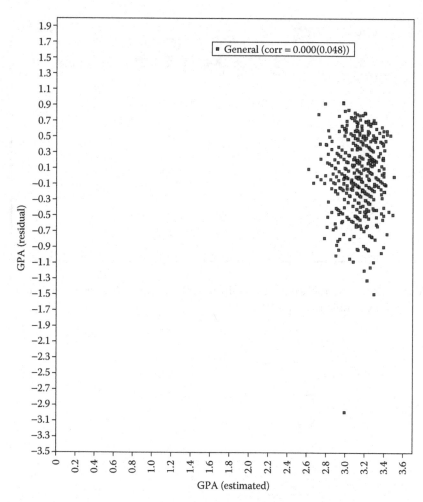

FIGURE 1.1
Homogeneity plot for regression model predicting GPA from `CTA_tot` and `BStotal`.

as is assumed in regression analysis. In order to check the normality of residuals assumption, Q–Q plots (quantile quantile plots) are typically used. We can obtain the Q–Q plot by first clicking on ⬚ , and then selecting Histograms (sample values, estimated values) in the following window.

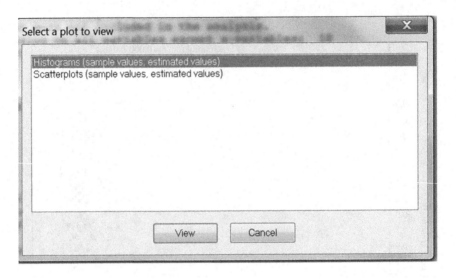

This yields the Histograms menu window. We will want to select GPA (residual) as the variable to be plotted.

In order to obtain the Q–Q plot, we click on the Display properties tab, and select QQ plot.

When we click OK, the plot will appear (Figure 1.3).

Interpretation of the Q–Q plot is quite simple. Essentially, the graph displays the data as it actually is on the x-axis and the data as it would be if it were normally distributed on the y-axis. The individual data points are represented in R by the black circles, and the solid line represents the data when they conform perfectly to the normal distribution. Therefore, the closer the observed data (circles) are to the solid line, the more closely the data conforms to the normal distribution. In this example, the data appear to follow the normal distribution fairly closely.

We can also obtain a histogram of the residuals by following a similar command sequence as that for the Q–Q plot (Figure 1.4). However, under the Display properties tab, we would select Standard histogram.

If we would like the normal density to be superimposed on the histogram, then we would select Histogram/density plot (Figure 1.5).

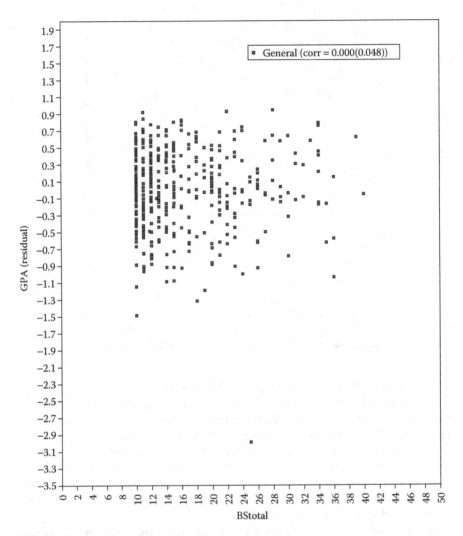

FIGURE 1.2
Diagnostic residuals plots for the regression model predicting GPA from CTA_tot and
BStotal.   (*Continued*)

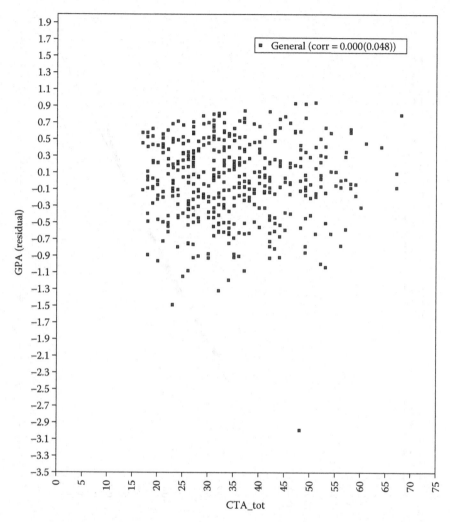

FIGURE 1.2 (Continued)
Diagnostic residuals plots for the regression model predicting GPA from CTA_tot and BStotal.

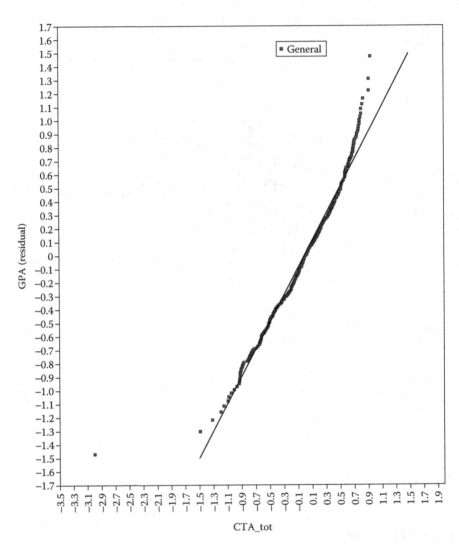

FIGURE 1.3
Q–Q plot for residuals of Model 1.1.

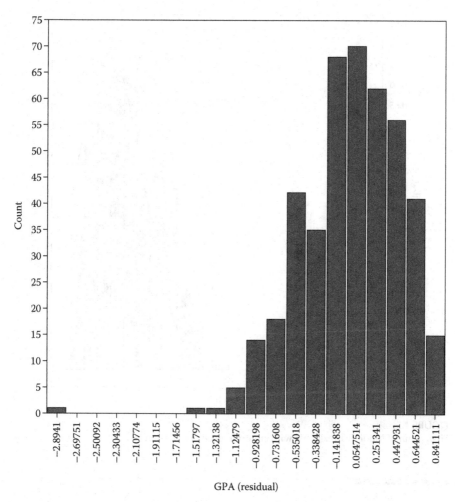

FIGURE 1.4
Histogram of Model 1.1 residuals.

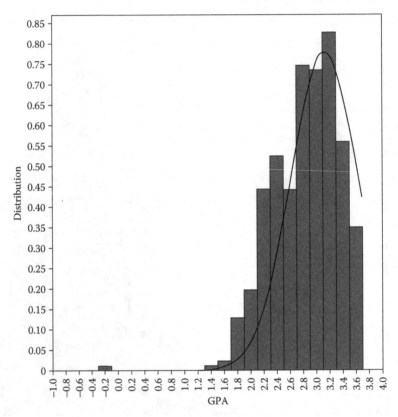

FIGURE 1.5
Histogram of Model 1.1 residuals with superimposed normal density.

Summary

This chapter introduced the reader to the basics of linear modeling using Mplus. This treatment was purposely limited, as there are a number of good resources available on this subject, and it is not the main focus of this book. However, many of the core concepts presented here for the GLM apply to multilevel modeling as well, and thus are of key importance as we move into these more complex analyses. In addition, much of the syntactical framework presented here will reappear in subsequent chapters. In particular, readers should leave this chapter comfortable with interpretation of coefficients in linear models, as well as the concept of variance explained in an outcome variable. We would encourage you to return to this chapter frequently as needed in order to reinforce these basic concepts. Next, in Chapter 2, we will turn our attention to the conceptual underpinnings of multilevel modeling before delving into their estimation in subsequent chapters.

2

---

*An Introduction to Multilevel
Data Structure*

Nested Data and Cluster Sampling Designs

In Chapter 1, we considered the standard linear model that underlies such common statistical methods as regression and analysis of variance (ANOVA; i.e., the general linear model). As we noted, this model rests on several primary assumptions regarding the nature of the data in the population. Of particular importance in the context of multilevel modeling is the assumption of independently distributed error terms for the individual observations within the sample. This assumption essentially means that there are no relationships among individuals in the sample for the dependent variable, *once the independent variables in the analysis are accounted for.* In the example we described in Chapter 1, this assumption was indeed met, as the individuals in the sample were selected randomly from the general population. Therefore, there was nothing linking their dependent variable values other than the independent variables included in the linear model. However, in many cases, the method used for selecting the sample does create the correlated responses among individuals. For example, a researcher interested in the impact of a new teaching method on student achievement might randomly select schools for placement in either a treatment or a control group. If school A is placed into the treatment condition, all students within the school will also be in the treatment condition—this is a cluster randomized design, in that the clusters and not the individuals are assigned to a specific group. Furthermore, it would be reasonable to assume that the school itself, above and beyond the treatment condition, would have an impact on the performance of the students. This impact would manifest itself as correlations in achievement test scores among individuals attending that school. Thus, if we were to use a simple one-way ANOVA to compare the achievement test means for the treatment and control groups with such cluster sampled data, we would likely be violating the assumption of independent errors because a factor beyond treatment condition (in this case the school) would have an additional impact on the outcome variable.

We typically refer to the data structure described previously as nested, meaning that individual data points at one level (e.g., student) appear in only one level of a higher level variable such as school. Thus, students are nested within school. Such designs can be contrasted with a crossed data structure whereby individuals at the first level appear in multiple levels of the second variable. In our example, students might be crossed with after school organizations if they are allowed to participate in more than one. For example, a given student might be on the basketball team as well as in the band. The focus of this book is almost exclusively on nested designs, which give rise to multilevel data. Other examples of nested designs might include a survey of job satisfaction for employees from multiple departments within a large business organization. In this case, each employee works within only a single division in the company, making leading to a nested design. Furthermore, it seems reasonable to assume that employees working within the same division will have correlated responses on the satisfaction survey, as much of their view regarding the job would be based exclusively upon experiences within their division. For a third such example, consider the situation in which clients of several psychotherapists working in a clinic are asked to rate the quality of each of their therapy sessions. In this instance, there exist three levels in the data: (1) time in the form of individual therapy session, (2) client, and (3) therapist. Thus, session is nested in client, who in turn is nested within therapist. All of this data structure would be expected to lead to correlated scores on a therapy rating instrument.

Intraclass Correlation

In cases where individuals are clustered or nested within a higher level unit (e.g., classrooms, schools, school districts), it is possible to estimate the correlation among individuals' scores within the cluster/nested structure using the intraclass correlation (denoted ρ_I in the population). The ρ_I is a measure of the proportion of variation in the outcome variable that occurs between groups versus the total variation present and ranges from 0 (no variance between clusters) to 1 (variance between clusters but no within cluster variance). ρ_I can also be conceptualized as the correlation for the dependent measure for two individuals randomly selected from the same cluster. It can be expressed as

$$\rho_I = \frac{\tau^2}{\tau^2 + \sigma^2} \qquad (2.1)$$

where:
  τ^2 is the population variance between clusters
  σ^2 is the population variance with clusters

Higher values of ρ_I indicate that a greater share of the total variation in the outcome measure is associated with cluster membership; that is, there is a relatively strong relationship among the scores for two individuals from the same cluster. Another way to frame this issue is that individuals within the same cluster (e.g., school) are more alike on the measured variable than they are like those in other clusters.

It is possible to estimate τ^2 and σ^2 using sample data, and thus is also possible to estimate ρ_I. Those familiar with ANOVA will recognize these estimates as being related (though not identical) to the sum of squared terms. The sample estimate for variation within clusters is simply

$$\hat{\sigma}^2 = \frac{\sum_{j=1}^{C}(n_j - 1)S_j^2}{N - C} \tag{2.2}$$

where:
S_j^2 is the variance within cluster $j = \sum_{i=1}^{n_j}(y_{ij} - \bar{y}_j)/(n_j - 1)$
n_j is the sample size for cluster j
N is the total sample size
C is the total number of clusters

In other words, σ^2 is simply the weighted average of within cluster variances.

Estimation of τ^2 involves a few more steps, but is not much more complex than what we have seen for σ^2. In order to obtain the sample estimate for variation between clusters, $\hat{\tau}^2$, we must first calculate the weighted between-cluster variance.

$$\hat{S}_B^2 = \frac{\sum_{j=1}^{C} n_j(\bar{y}_j - \bar{y})^2}{\tilde{n}(C - 1)} \tag{2.3}$$

where:
$\bar{y}_j$ is the mean on response variable for cluster j
$\bar{y}$ is the overall mean on response variable

$$\tilde{n} = \frac{1}{C - 1}\left[N - \frac{\sum_{j=1}^{C} n_j^2}{N} \right]$$

We cannot use S_B^2 as a direct estimate of τ^2 because it is impacted by the random variation among subjects within the same clusters. Therefore, in order to remove this random fluctuation, we will estimate the population between-cluster variance as

$$\hat{\tau}^2 = S_B^2 - \frac{\hat{\sigma}^2}{\tilde{n}} \tag{2.4}$$

Using these variance estimates, we can in turn calculate the sample estimate of ρ_I:

$$\hat{\rho}_I = \frac{\hat{\tau}^2}{\hat{\tau}^2 + \hat{\sigma}^2} \tag{2.5}$$

Note that the above equation assumes that the clusters are of equal size. Clearly, such will not always be the case, in which case this equation will not hold. However, the purpose for its inclusion here is to demonstrate the principal underlying the estimation of ρ_I, which holds even as the equation might change.

In order to illustrate estimation of ρ_I, let us consider the following dataset: achievement test data were collected from 10,903 third-grade examinees nested within 160 schools. School sizes ranges from 11 to 143, with a mean size of 68.14. In this case, we will focus on the reading achievement test score and will use data from only five of the schools, in order to make the calculations by hand easy to follow. First, we will estimate $\hat{\sigma}^2$. To do so, we must estimate the variance in scores within each school. These values appear in Table 2.1.

Using these variances and sample sizes, we can calculate $\hat{\sigma}^2$ as

$$\hat{\sigma}^2 = \frac{\sum_{j=1}^{C}(n_j - 1)S_j^2}{N - C}$$

$$= \frac{(58-1)5.3 + (29-1)1.5 + (64-1)2.9 + (39-1)6.1 + (88-1)3.4}{278-5}$$

$$= \frac{302.1 + 42 + 182.7 + 231.8 + 295.8}{273}$$

$$= \frac{1054.4}{273}$$

$$= 3.9$$

TABLE 2.1

School Size, Mean, and Variance of
Reading Achievement Test

School	N	Mean	Variance
767	58	3.952	5.298
785	29	3.331	1.524
789	64	4.363	2.957
815	39	4.500	6.088
981	88	4.236	3.362
Total	278	4.149	3.916

TABLE 2.2

Multilevel Model Parameter Estimates

School	Intercept	U_{0j}	Slope	U_{1j}
1	1.230	−1.129	0.552	0.177
2	2.673	0.314	0.199	−0.176
3	2.707	0.348	0.376	0.001
4	2.867	0.508	0.336	−0.039
5	2.319	−0.040	0.411	0.036
Overall	2.359		0.375	

The school means, which are needed in order to calculate S_B^2, appear in Table 2.2 as well. First, we must calculate $\tilde{n}$:

$$\tilde{n} = \frac{1}{C-1}\left(N - \frac{\sum_{j=1}^{C} n_j^2}{N} \right)$$

$$= \frac{1}{5-1}\left(278 - \frac{58^2 + 29^2 + 64^2 + 39^2 + 88^2}{278} \right)$$

$$= \frac{1}{4}\left(278 - 63.2 \right)$$

$$= 53.7$$

Using this value, we can then calculate S_B^2 for the five schools in our small sample using Equation 2.3:

$$\frac{\left[\begin{array}{c} 58(3.952-4.149)^2 + 29(3.331-4.149)^2 + 64(4.363-4.149)^2 \\ + 39(4.500-4.149)^2 + 88(4.236-4.149) \end{array} \right]}{53.7(5-1)}$$

$$= \frac{2.251 + 19.405 + 2.931 + 4.805 + 0.666}{214.8} = \frac{30.057}{214.800} = 0.140$$

We can now estimate the population between cluster variance, τ^2, using Equation 2.4:

$$0.140 - \frac{3.9}{53.7} = 0.140 - 0.073 = 0.067$$

We have now calculated all of the parts that we need to estimate ρ_I for the population:

$$\hat{\rho}_I = \frac{0.067}{0.067 + 3.9} = 0.017$$

This result indicates that there is very little correlation of examinees' test scores within the schools. We can also interpret this value as the proportion of variation in the test scores that is accounted for by the schools.

Given that $\hat{\rho}_I$ is a sample estimate, we know that it is subject to sampling variation, which can be estimated with a standard error as in the equation:

$$S_{\rho_I} = (1 - \rho_I)(1 + (n - 1)\rho_I)\sqrt{\frac{2}{n(n-1)(N-1)}} \tag{2.6}$$

The terms in the above equation are as defined previously, and the assumption is that all clusters are of equal size. As noted earlier in the chapter, this latter condition is not a requirement, however, and an alternative formulation exists for cases in which it does not hold. However, Equation 2.6 provides sufficient insight for our purposes into the estimation of the standard error of the intraclass correlation (ICC).

The ICC is an important tool in multilevel modeling, in large part because it is an indicator of the degree to which the multilevel data structure might impact the outcome variable of interest. Larger values of the ICC are indicative of a greater impact of clustering. Thus, as the ICC increases in value, we must be more cognizant of employing multilevel modeling strategies in our data analysis. In section "Pitfalls of Ignoring Multilevel Data Structure", we will discuss the problems associated with ignoring this multilevel structure, before we turn our attention to methods for dealing with it directly.

Pitfalls of Ignoring Multilevel Data Structure

When researchers apply standard statistical methods to multilevel data, such as the regression model described in Chapter 1, the assumption of independent errors is violated. For example, if we have achievement test scores from a sample of students who attend several different schools, it would be reasonable to believe that those attending the same school will have scores that are more highly correlated with one another than they are with scores from students attending other schools. This within-school correlation would be due, for example, to having a common set of teachers, a common teaching curriculum, coming from a common community, a single set of administrative policies, among numerous other reasons. The within-school correlation will result in an inappropriate estimate of the of the standard errors for the model parameters, which in turn will lead to errors of statistical inference, such as p-values smaller than they really should be and the resulting rejection of null effects above the stated Type I error rate, regarding the parameters. Recalling our discussion in Chapter 1, the test statistic for the null hypothesis of no relationship between the independent and dependent variable is simply the regression coefficient divided by the standard error. If the standard error is underestimated, this will lead to an overestimation of the test statistic, and therefore statistical significance for the parameter in cases where it should not be (i.e., a Type I error at a higher rate than specified). Indeed, the underestimation of the standard error will occur unless τ^2 is equal to 0.

In addition to the underestimation of the standard error, another problem with ignoring the multilevel structure of data is that we may miss important relationships involving each level in the data. Recall that in our example, there are two levels of sampling: students (level 1) are nested in schools (level 2). Specifically, by *not* including information about the school, for example, we may well miss important variables at the school level that help to explain performance at the examinee level. Therefore, beyond the known problem with misestimating standard errors, we also proffer an incorrect model for understanding the outcome variable of interest. In the context of multilevel linear models (MLMs), inclusion of variables at each level is relatively simple, as are interactions among variables at different levels. This greater model complexity in turn may lead to greater understanding of the phenomenon under study.

Multilevel Linear Models

In the section "Random Intercepts," we will review some of the core ideas that underlie MLMs. Our goal is to familiarize the reader with terms that will repeat themselves throughout the book, and to do so in a relatively nontechnical fashion. We will first focus on the difference between random and fixed effects, after which we will discuss the basics of parameter estimation, focusing on the two most commonly used methods, maximum likelihood and restricted maximum likelihood, and conclude with a review of assumptions underlying MLMs, and overview of how they are most frequently used, with examples. Included in this section, we will also address the issue of centering and explain why it is an important concept in MLM. After reading the rest of this chapter, the reader will have sufficient technical background on MLMs to begin using the R software package for fitting MLMs of various types.

Random Intercept

As we transition from the one-level regression framework of Chapter 1 to the MLM context, let us first revisit the basic simple linear regression model of

Equation 1.1, $y = \beta_0 + \beta_1 x + \varepsilon$. Here, the dependent variable y is expressed as a function of an independent variable, x, multiplied by a slope coefficient, β_1, an intercept, β_0, and random variation from subject to subject, ε. We defined the intercept as the conditional mean of y when the value of x is 0. In the context of a single-level regression model such as this, there is one intercept that is common to all individuals in the population of interest. However, when individuals are clustered together in some fashion (e.g., within classrooms, schools, organizational units within a company), there will potentially be a separate intercept for each of these clusters; that is, there may be different means for the dependent variable for $x = 0$ across the different clusters. We say *potentially* here because if there is in fact no cluster effect, then the single intercept model of 1.1 will suffice. In practice, assessing whether there are different means across the clusters is an empirical question, which we describe next. It should also be noted that in this discussion we are considering only the case where the intercept is cluster specific, but it is also possible for β_1 to vary by group as well, or even other coefficients from more complicated models.

Allowing for group-specific intercepts and slopes leads to the following notation commonly used for the level 1 (micro level) model in multilevel modeling:

$$y_{ij} = \beta_{0j} + \beta_{1j} x + \varepsilon_{ij} \qquad (2.7)$$

where the subscripts ij refer to the ith individual in the jth cluster.

As we continue our discussion of multilevel modeling notation and structure, we will begin with the most basic multilevel model: predicting the outcome from just an intercept which we will allow to vary randomly for each group.

$$y_{ij} = \beta_{0j} + \varepsilon_{ij} \qquad (2.8)$$

Allowing the intercept to differ across clusters, as in Equation 2.8, leads to the random intercept which we express as

$$\beta_{0j} = \gamma_{00} + U_{0j} \qquad (2.9)$$

In this framework, γ_{00} represents an average or general intercept value that holds across clusters, whereas U_{0j} is a group-specific effect on the intercept. We can think of γ_{00} as a fixed effect because it remains constant across all clusters, and U_{0j} is a random effect because it varies from cluster to cluster. Therefore, for an MLM we are interested not only in some general mean value for y when x is 0 for all individuals in the population (γ_{00}) but also in the deviation between the overall mean and the cluster specific effects for the intercept (U_{0j}). If we go on to assume that the clusters are a random sample from the population of all such clusters, then we can treat the

U_{0j} as a kind of residual effect on y_{ij}, very similar to how we think of ε. In that case, U_{0j} is assumed to be drawn randomly from a population with a mean of 0 (recall U_{0j} is a deviation from the fixed effect) and a variance, τ^2. Furthermore, we assume that τ^2 and σ^2, the variance of ε, are uncorrelated. We have already discussed τ^2 and its role in calculating $\hat{\rho}_I$. In addition, τ^2 can also be viewed as the impact of the cluster on the dependent variable, and therefore, testing it for statistical significance is equivalent to testing the null hypothesis that cluster (e.g., school) has no impact on the dependent variable. If we substitute the two components of the random intercept into the regression model, we get

$$y = \gamma_{00} + U_{0j} + \beta_1 x + \varepsilon \tag{2.10}$$

The above equation is termed the full or composite model in which the multiple levels are combined into a unified equation.

Often in MLM, we begin our analysis of a dataset with this simple random intercept model, known as the null model, which takes the form:

$$y_{ij} = \gamma_{00} + U_{0j} + \varepsilon_{ij} \tag{2.11}$$

Although the null model does not provide information regarding the impact of specific independent variables on the dependent, it does yield important information regarding how variation in y is partitioned between variance among the individuals σ^2 and variance among the clusters τ^2. The total variance of y is simply the sum of σ^2 and τ^2. In addition, as we have already seen, these values can be used to estimate ρ_I. The null model, as will be seen in later sections, is also used as a baseline for model building and comparison.

Random Slopes

It is a simple matter to expand the random intercept model in Equation 2.9 to accommodate one or more independent predictor variables. As an example, if we add a single predictor (x_{ij}) at the individual level (level 1) to the model, we obtain

$$y_{ij} = \gamma_{00} + \gamma_{10} x_{ij} + U_{0j} + \varepsilon_{ij} \tag{2.12}$$

This model can also be expressed in two separate levels as follows:

$$\text{Level 1: } y_{ij} = \beta_{0j} + \beta_{1j} x + \varepsilon_{ij} \tag{2.13}$$

$$\text{Level 2: } \beta_{0j} = \gamma_{00} + U_{0j} \tag{2.14}$$

$$\beta_{1j} = \gamma_{10} \tag{2.15}$$

This model now includes the predictor and the slope relating it to the dependent variable, γ_{10}, which we acknowledge as being at level 1 by the subscript 10. We interpret γ_{10} in the same way that we did β_1 in the linear regression model; that is, a measure of the impact on y of a 1 unit change in x. In addition, we can estimate ρ_I exactly as before, though now it reflects the correlation between individuals from the same cluster after controlling for the independent variable, x. In this model, both γ_{10} and γ_{00} are fixed effects, whereas σ^2 and τ^2 remain random.

One implication of the model in Equation 2.12 is that the dependent variable is impacted by variation among individuals (σ^2), variation among clusters (τ^2), an overall mean common to all clusters (γ_{00}), and the impact of the independent variable as measured by γ_{10}, which is also common to all clusters. In practice, there is no reason that the impact of x on y would need to be common for all clusters, however. In other words, it is entirely possible that rather than having a single γ_{10} common to all clusters, there is actually a unique effect for the cluster of $\gamma_{10} + U_{1j}$, where γ_{10} is the average relationship of x with y across clusters, and U_{1j} is the cluster-specific variation of the relationship between the two variables. This cluster-specific effect is assumed to have a mean of 0 and to vary randomly around γ_{10}. The random slopes model is

$$y_{ij} = \gamma_{00} + \gamma_{10}x_{ij} + U_{0j} + U_{1j}x_{ij} + \varepsilon_{ij} \qquad (2.16)$$

Written in this way, we have separated the model into its fixed ($\gamma_{00} + \gamma_{10}x_{ij}$) and random ($U_{0j} + U_{1j}x_{ij} + \varepsilon_{ij}$) components. Model 2.16 simply states that there is an interaction between cluster and x, such that the relationship of x and y is not constant across clusters.

Heretofore, we have discussed only one source of between-group variation, which we have expressed as τ^2, and which is the variation among clusters in the intercept. However, Model 2.16 adds a second such source of between-group variance in the form of U_{1j}, which is cluster variation on the slope relating the independent and dependent variables. In order to differentiate between these two sources of between-group variance, we now denote the variance of U_{0j} as τ_0^2 and the variance of U_{1j} as τ_1^2. Furthermore, within clusters we expect U_{1j} and U_{0j} to have a covariance of τ_{01}. However, across different clusters, these terms should be independent of one another, and in all cases, it is assumed that ε remains independent of all other model terms. In practice, if we find that τ_1^2 is not 0, we must be careful in describing the relationship between the independent and dependent variables, as it is not the same for all clusters. We will revisit this idea in subsequent chapters. For the moment, however, it is most important to recognize that variation in the dependent variable, y, can be explained by several sources, some fixed and others random. In practice, we will most likely be interested in estimating all of these sources of variability in a single model.

As a means for further understanding the MLM, let us consider a simple example using the five schools described earlier. In this context, we are interested in treating a reading achievement test score as the dependent variable and a vocabulary achievement test score as the independent variable. Remember that students are nested within schools so that a simple regression analysis will not be appropriate. In order to understand what is being estimated in the context of MLM, we can obtain separate intercept and slope estimates for each school, which appear in Table 2.2.

Given that the schools are of the same sample size, the estimate of γ_{00}, the average intercept value is 2.359, and the estimate of the average slope value, γ_{10}, is 0.375. Notice that for both parameters, the school values deviate from these means. For example, the intercept for school 1 is 1.230. The difference between this value and 2.359, −1.129, is U_{0j} for that school. Similarly, the difference between the average slope value of 0.375 and the slope for school 1, 0.552, is 0.177, which is U_{1j} for this school. Table 2.2 includes U_{0j} and U_{1j} values for each of the schools. The differences in slopes also provide information regarding the relationship between vocabulary and reading test scores. For all of the schools, this relationship was positive, meaning that students who scored higher on vocabulary also scored higher on reading. However, the strength of this relationship was weaker for school 2 than for school 1, as an example.

Given the values in Table 2.2, it is also possible to estimate the variances associated with U_{1j} and U_{0j}, τ_1^2 and τ_0^2, respectively. Again, because the schools in this example had the same number of students, the calculation of these variances is a straightforward matter, using

$$\frac{\sum\left(U_{1j}-\overline{U}_1\right)^2}{J-1} \tag{2.17}$$

for the slopes, and an analogous equation for the intercept random variance. Doing so, we obtain $\tau_0^2 = 0.439$ and $\tau_1^2 = 0.016$. In other words, much more of the variance in the dependent variable is accounted for by variation in the intercepts at the school level than is accounted for by variation in the slopes. Another way to think of this result is that the schools exhibited greater differences among one another in the mean level of achievement compared to differences in the impact of x on y.

The actual practice of obtaining these variance estimates using the R environment for statistical computing and graphics and interpreting their meaning are subjects for the coming chapters. Before discussing the practical nuts and bolts of conducting this analysis, we will first examine the basics for how parameters are estimated in the MLM framework using maximum likelihood and REML algorithms. Although similar in spirit to the simple calculations demonstrated previously, they are different in practice and will yield somewhat different results than those we would obtain using least squares, as before. Prior to this discussion, however, there is one more issue that warrants our attention as we consider the practice of MLM, namely, variable centering.

Centering

Centering simply refers to the practice of subtracting the mean of a variable from each individual value. This implies the mean for the sample of the centered variables is 0 and implies that each individual's (centered) score represents a deviation from the mean, rather than whatever meaning its raw value might have. In the context of regression, centering is commonly used, for example, to reduce collinearity caused by including an interaction term in a regression model. If the raw scores of the independent variables are used to calculate the interaction, and then both the main effects and the interaction terms are included in the subsequent analysis, it is very likely that collinearity will cause problems in the standard errors of the model parameters. Centering is a way to help avoid such problems (e.g., Iversen, 1991). Such issues are also important to consider in MLM, in which interactions are frequently employed. In addition, centering is also a useful tool for avoiding collinearity caused by highly correlated random intercepts and slopes in MLMs (Wooldridge, 2004). Finally, it provides a potential advantage in terms of interpretation of results. Remember from our discussion in Chapter 1 that the intercept is the value of the dependent variable when the independent variable is set equal to 0. In many applications, the independent variable cannot reasonably be 0 (e.g., a measure of vocabulary), however, which essentially renders the intercept as a necessary value for fitting the regression line but not one that has a readily interpretable value. However, when x has been centered, the intercept takes on the value of the dependent variable when the independent is at its mean. This is a much more useful interpretation for researchers in many situations, and yet another reason why centering is an important aspect of modeling, particularly in the multilevel context.

Probably, the most common approach to centering is to calculate the difference between each individual's score and the overall, or grand mean across the entire sample. This *grand mean centering* is certainly the most commonly used in practice (Bickel, 2007). It is not, however, the only manner in which data can be centered. An alternative approach, known as *group mean centering*, is to calculate the difference between each individual score and the mean of the cluster to which they belong. In our school example, grand mean centering would involve calculating the difference between each score and the overall mean across schools, whereas group mean centering would lead the researcher to calculate the difference between each score and the mean for their school. Although there is some disagreement in the literature regarding which approach might be best at reducing the harmful effects of collinearity (Raudenbush & Bryk, 2002; Snijders & Bosker, 1999), researchers have demonstrated that in most cases either will work well in this regard (Kreft, de Leeuw, & Aiken, 1995). Therefore, the choice of which approach to use must be made on substantive grounds regarding the nature of the relationship between x and y. By using grand mean centering, we are implicitly comparing individuals to one another (in the form of the overall mean) across the entire sample. However, when using group mean centering, we are placing each individual in relative position on x within their cluster. Thus, in our school example, using the group mean centered values of vocabulary in the analysis would mean that we are investigating the relationship between one's relative vocabulary score in his or her school and his or her reading score. In contrast, the use of grand mean centering would examine the relationship between one's relative standing in the sample as a whole on vocabulary and the reading score. This latter interpretation would be equivalent conceptually (though not mathematically) to using the raw score, whereas the group mean centering would not. Throughout the rest of this book, we will use grand mean centering by default, per recommendations by Hox (2002), among others. At times, however, we will also demonstrate the use of group mean centering in order to illustrate how it provides different results, and for applications in which interpretation of the impact of an individual's relative standing in their cluster might be more useful than their relative standing in the sample as a whole.

Basics of Parameter Estimation with MLMs

Heretofore, when we have discussed estimation of model parameters, it has been in the context of least squares, which serves as the underpinnings of ordinary least squares (OLS) and related linear models. However, as we move from these fairly simple applications to more complex models, OLS is not typically the optimal approach to use for parameter estimation. Instead, we will rely on maximum likelihood estimation (MLE) and restricted maximum likelihood (REML). In the following sections, "Maximum Likelihood Estimation," and "Restricted Likelihood Estimation," we review these approaches to estimation from a conceptual basis, focusing on the generalities of how they work, what they assume about the data, and how they differ from one another. For the technical details, we refer the interested reader to the work of Raudenbush and Bryk (2002) or de Leeuw and Meijer (2008), both of which provide excellent resources for those desiring a more in depth coverage of these methods. Our purpose here is to provide the reader with a conceptual understanding that will aid in their understanding of application of MLM in practice.

Maximum Likelihood Estimation

MLE has as its primary goal the estimation of population model parameters that maximize the likelihood of our obtaining the sample that we in fact obtained. In other words, the estimated parameter values should maximize the likelihood of our particular sample. From a practical perspective, identifying such sample values takes place through the comparison of the observed data with that predicted by the model associated with the parameter values. The closer the observed and predicted values are to one another, the greater the likelihood that the observed data arose from a population with parameters close to those used to generate the predicted values. In practice, MLE is an iterative methodology in which the algorithm searches for those parameter values that will maximize the likelihood of the observed data (i.e., produce predicted values that are as close as possible to the observed), and as such can be computationally intensive, particularly for complex models and large samples.

Restricted Maximum Likelihood Estimation

There exists a variant of MLE, REML, that has been shown to be more accurate with regard to the estimation of variance parameters than is MLE (Kreft & de Leeuw, 1998). In particular, the two methods differ with respect to how degrees of freedom are calculated in the estimation of variances. As a simple example, the sample variance is typically calculated by dividing the sum of squared differences between individual values and the mean by the number of observations minus 1, so as to have an unbiased estimate. This is a REML estimate of variance. In contrast, the MLE variance is calculated by dividing the sum of squared differences by the total sample size, leading to a smaller variance estimate than REML and, in fact, one that is biased in finite samples. In the context of multilevel modeling, REML takes into account the number of parameters being estimated in the model when determining the appropriate degrees of freedom for the estimation of the random components such as the parameter variances described previously. In contrast, MLE does not account for these, leading to an underestimate of the variances that does not occur with REML. For this reason, REML is generally the preferred method for estimating multilevel models, though for testing variance parameters (or any random effect) it is necessary to use MLE (Snijders & Bosker, 1999). It should be noted that as the number of level 2 clusters increases, the difference in value for MLE and REML estimates becomes very small (Snijders & Bosker, 1999).

Assumptions Underlying MLMs

As with any statistical model, the appropriate use of MLMs requires that several assumptions about the data hold true. If these assumptions are not met, the model parameter estimates may not be trustworthy, just as would be the case with standard linear regression that was reviewed in Chapter 1. Indeed, although they differ somewhat from the assumptions for the single-level models, the assumptions underlying MLM are akin to those for the simpler models. In this section, we provide an introduction to these assumptions and their implications for researchers using MLMs, and in subsequent chapters we describe methods for checking the validity of these assumptions for a given set of data.

First, we assume that the level 2 residuals are independent between clusters. In other words, there is an assumption that the random intercept and slope(s) at level 2 are independent of one another across clusters. Second, the level 2 intercepts and coefficients are assumed to be independent of the level 1 residuals; that is, the errors for the cluster level estimates are unrelated to errors at the individual level. Third, the level 1 residuals are normally distributed and have a constant variance. This assumption is very similar to the one we make about residuals in the standard linear regression model. Fourth, the level 2 intercept and slope(s) have a multivariate normal distribution with a constant covariance matrix. Each of these assumptions can be directly assessed for a sample, as we shall see in forthcoming chapters. Indeed, the methods for checking the MLM assumptions are not very different than those for checking the regression model that we used in Chapter 1.

Overview of Two-Level MLMs

To this point, we have described the specific terms that make up the MLM, including the level 1 and level 2 random effects and residuals. We will close out this chapter introducing the MLM by considering an example of two- and three-level MLMs, and the use of MLM with longitudinal data. This discussion should prepare the reader for subsequent chapters in which we consider the application of R to the estimation of specific MLMs. First, let us consider the two-level MLM, parts of which we have already described previously in the chapter.

Previously, in Equation 2.16, we considered the random slopes model $y_{ij} = \gamma_{00} + \gamma_{10}x_{ij} + U_{0j} + U_{1j}x_{ij} + \varepsilon_{ij}$ in which the dependent variable, y_{ij} (reading achievement), was a function of an independent variable x_{ij} (vocabulary test score), as well as random error at both the examinee and school levels. We can extend this model a bit further by including multiple independent variables at both level 1 (examinee) and level 2 (school). Thus, for example, in addition to ascertaining the relationship between an individual's vocabulary and reading scores, we can also determine the degree to which the average vocabulary score at the school as a whole is related to an individual's reading score. This model would essentially have two parts, one explaining the relationship between the individual-level vocabulary (x_{ij}) and reading, and the other explaining the coefficients at level 1 as a function of the level 2 predictor, average vocabulary score (z_j). The two parts of this model are expressed as follows:

$$\text{Level 1: } y_{ij} = \beta_{0j} + \beta_{1j}x_{ij} + \varepsilon_{ij} \tag{2.18}$$

and

$$\text{Level 2: } \beta_{hj} = \gamma_{h0} + \gamma_{h1}z_j + U_{hj} \tag{2.19}$$

The additional piece of the equation in Equation 2.19 is $\gamma_{h1}z_j$, which represents the slope for (γ_{h1}), and the value of the average vocabulary score for the school (z_j). In other words, the mean school performance is related directly to the coefficient linking the individual vocabulary score to the individual reading score. For our specific example, we can combine Equations 2.18 and 2.19 in order to obtain a single equation for the two-level MLM.

$$y_{ij} = \gamma_{00} + \gamma_{10}x_{ij} + \gamma_{01}z_j + \gamma_{1001}x_{ij}z_j + U_{0j} + U_{1j}x_{ij} + \varepsilon_{ij} \tag{2.20}$$

Each of these model terms has been defined previously in the chapter: γ_{00} is the intercept or the grand mean for the model, γ_{10} is the fixed effect of variable x (vocabulary) on the outcome, U_{0j} represents the random variation for the intercept across groups, and U_{1j} represents the random variation for the slope across groups. The additional pieces of Equation 2.13 are γ_{01} and γ_{11}. γ_{01} represents the fixed effect of level 2 variable z (average vocabulary) on the outcome, whereas γ_{11} represents the slope for, and value of the average vocabulary score for the school. The new term in model 2.20 is the cross-level interaction, $\gamma_{1001}x_{ij}z_j$. As the name implies, the cross-level interaction is simply the interaction between the level 1 and level 2 predictors. In this context, it is the interaction between an individual's vocabulary score and the mean vocabulary score for his or her school. The coefficient for this interaction term, γ_{1001}, assesses the extent to which the relationship between an examinee's vocabulary score is moderated by the mean for the school that they attend. A large significant value for this coefficient would indicate that the relationship between a person's vocabulary test score and his or her overall reading achievement is dependent on the level of vocabulary achievement at his or her school.

Overview of Three-Level MLMs

With MLMs, it is entirely possible to have three or more levels of data structure. It should be noted that in actual practice four-level and higher models are rare, however. For our reading achievement data, where the second level was school, the third level might be the district in which the school resides. In that case, we would have multiple equations to consider when expressing the relationship between vocabulary and reading achievement scores, starting at the individual level:

$$y_{ijk} = \beta_{0jk} + \beta_{1jk}x_{ijk} + \varepsilon_{ijk} \tag{2.21}$$

Here, the subscript k represents the level 3 cluster to which the individual belongs. Prior to formulating the rest of the model, we must evaluate if the slopes and intercepts are random at both levels 2 and 3, or only at level 1, for example. This decision should always be based on the theory surrounding the research questions, what is expected in the population, and what is revealed in the empirical data. We will proceed with the remainder of this discussion under the assumption that the level 1 intercepts and slopes are random for both levels 2 and 3, in order to provide a complete description of the most complex model possible when three levels of data structure are present. When the level 1 coefficients are not random at both levels, the terms in the following models for which this randomness is not present would simply be removed. We will address this issue more specifically in Chapter 4, when we discuss the fitting of three-level models using R.

The level 2 and level 3 contributions to the MLM described in Equation 2.13 appear as follows:

$$\text{Level 2: } \beta_{0jk} = \gamma_{00k} + U_{0jk}$$

$$\beta_{1jk} = \gamma_{10k} + U_{1jk}$$

$$\text{Level 3: } \gamma_{00k} = \delta_{000} + V_{00k} \tag{2.22}$$

$$\gamma_{10k} = \delta_{100} + V_{10k}$$

We can then use a simple substitution to obtain the expression for the level 1 intercept and slope in terms of both level 2 and level 3 parameters.

$$\beta_{0jk} = \delta_{000} + V_{00k} + U_{0jk}$$

$$\beta_{1jk} = \delta_{100} + V_{10k} + U_{1jk} \tag{2.23}$$

In turn, these terms can be substituted into Equation 2.15 to provide the full three-level MLM.

$$y_{ijk} = \delta_{000} + V_{00k} + U_{0jk} + \left(\delta_{100} + V_{10k} + U_{1jk} \right) x_{ijk} + \varepsilon_{ijk} \tag{2.24}$$

There is an implicit assumption in this expression of 2.24 that there are no cross-level interactions, though these are certainly possible to model across all three levels, or for any pair of levels. In Equation 2.24, we are expressing individuals' scores on the reading achievement test as a function of random and fixed components from the school which they attend, the district in which the school resides, as well as their own vocabulary test score and random variation associated only with themselves. Though not present in Equation 2.24, it is also possible to include variables at both levels 2 and 3, similar to what we described for the two-level model structure.

Overview of Longitudinal Designs and Their Relationship to MLMs

Finally, we will say just a word about how longitudinal designs can be expressed as MLMs. Longitudinal research designs simply involve the collection of data from the same individuals at multiple points in time. For example, we may have reading achievement scores for examinees in the fall and spring of the school year. With such a design, we would be able to investigate issues around growth scores and change in achievement over time. Such models can be placed in the context of an MLM where the examinee is the level 2 (cluster) variable, and the individual test administration is at level 1. We would then simply apply the two-level model described previously, including whichever examinee level variables that are appropriate for explaining reading achievement. Similarly, if examinees are nested within schools, we would have a three-level model, with school at the third level, and could apply model 2.24, once again with whichever examinee- or school-level variables were pertinent to the research question. One unique aspect of fitting longitudinal data in the MLM context is that the error terms can potentially take specific forms that are not common in other applications of multilevel analysis. These error terms reflect the way in which measurements made over time are related to one another and are typically more complex than the basic error structure that we have described thus far. In Chapter 5, we will look at examples of fitting such longitudinal models with Mplus and focus much of our attention on these error structures, when each is appropriate, and how they are interpreted. In addition, such MLMs need not take a linear form, but can be adapted to fit quadratic, cubic, or other nonlinear trends over time. These issues will be further discussed in Chapter 5.

Summary

The goal of this chapter was to introduce the basic theoretical underpinnings of multilevel modeling, but not to provide an exhaustive technical discussion of these issues, as there are a number of useful sources available in this regard, which you will find among the references at the end of the text. However, what is given here should stand you in good stead as we move forward with multilevel modeling using Mplus. We recommend that while reading subsequent chapters, you make liberal use of the information provided here, in order to gain a more complete understanding of the output that we will be examining from Mplus. In particular, when interpreting output from Mplus, it may be very helpful for you to come back to this chapter for reminders on precisely what each model parameter means. In Chapters 3, 4, and 5, we will take the theoretical information from Chapter 2 and apply it to real datasets using Mplus. In Chapter 6, we will examine how these ideas can be applied to longitudinal data, and in Chapters 7 and 8, we will discuss multilevel modeling for categorical dependent variables. Chapters 9 and 10 focus on the application of Mplus to fitting multilevel latent variable models. Finally, in Chapter 11, we will diverge from the likelihood-based approaches described here and discuss multilevel modeling within the Bayesian framework, focusing on application, and learning when this method might be appropriate and when it might not.

3

Fitting Two-Level Models in Mplus

In Chapter 2, the multilevel modeling approach to analysis of nested data was introduced, along with relevant notation and definitions of random intercepts and coefficients. We will now devote Chapter 3 to the introduction of fitting multilevel models using Mplus software. After providing a brief discussion of the structure of the input commands for multilevel modeling, we will devote the remainder of the chapter to extended examples applying the principles that we introduced in Chapter 2 using Mplus.

Simple (Intercept Only) Multilevel Models

In order to demonstrate the use of Mplus for fitting multilevel models, let us return to the example that we introduced in Chapter 2. Specifically, a researcher is interested in determining the extent to which vocabulary scores can be used to predict general reading achievement. Students were nested within schools, so that standard linear regression models will not be appropriate. In this case, school is a random effect, whereas vocabulary scores are fixed. The first model that we will fit is the null model in which there is not an independent variable. This model is useful for obtaining the estimates of the residual and intercept variance when only the clustering by school is considered, as in Equation 2.11. The syntax for definition of a two-level null model can be found as follows:

```
TITLE: 3.0: Simple (Intercept Only) Multilevel Models.
The Null Model
DATA: FILE IS Achieve_ch3.csv;
VARIABLE: NAMES ARE corp school class classID gender
age geread gevocab clenroll senroll;
        USEVARIABLES ARE school geread;
        MISSING ARE .;
        CLUSTER = school;

ANALYSIS: TYPE = TWOLEVEL;
```

When defining multilevel models in Mplus, the USEVARIABLES statement defines all of the variables that will be used in the model. The CLUSTER statement is used to define the nesting structure of the data. Thus, in this example, the only variable defined in the USEVARIABLES statement is geread, and the CLUSTER variable is school because our data structure is students nested within schools. The ANALYSIS TYPE = TWOLEVEL statement is used to indicate that we are fitting a multilevel model. When running a null model, a MODEL statement is not required because there are no between or within cluster predictors to define. The output generated from this syntax can be found as follows:

```
Mplus VERSION 7.4
MUTHEN & MUTHEN
05/17/2016   2:19 PM

INPUT INSTRUCTIONS

   TITLE:  3.0: Simple (Intercept Only) Multilevel Models.
   The Null Model
   DATA: FILE IS Achieve_ch3.csv;
   VARIABLE: NAMES ARE corp school class classID gender age
   geread gevocab clenroll senroll;
           USEVARIABLES ARE school geread;
           MISSING ARE .;
           CLUSTER = school;

   ANALYSIS: TYPE = TWOLEVEL;



*** WARNING in MODEL command
   All variables are uncorrelated with all other variables in
   the model.
   Check that this is what is intended.
   1 WARNING(S) FOUND IN THE INPUT INSTRUCTIONS



3.0: Simple (Intercept Only) Multilevel Models.  The Null Model

SUMMARY OF ANALYSIS

Number of groups                                              1
Number of observations                                   10790

Number of dependent variables                                1
Number of independent variables                              0
Number of continuous latent variables                        0

Observed dependent variables

   Continuous
    GEREAD
```

```
Variables with special functions

    Cluster variable        SCHOOL

Estimator                                                MLR
Information matrix                                   OBSERVED
Maximum number of iterations                             100
Convergence criterion                            0.100D-05
Maximum number of EM iterations                         500
Convergence criteria for the EM algorithm
  Loglikelihood change                           0.100D-02
  Relative loglikelihood change                  0.100D-05
  Derivative                                     0.100D-03
Minimum variance                                 0.100D-03
Maximum number of steepest descent iterations            20
Maximum number of iterations for H1                    2000
Convergence criterion for H1                     0.100D-03
Optimization algorithm                                   EMA

Input data file(s)
  Achieve_ch3.csv
Input data format  FREE


SUMMARY OF DATA

      Number of missing data patterns          1
      Number of clusters                     163

      Average cluster size        66.196

      Estimated Intraclass Correlations for the Y Variables

                  Intraclass
      Variable   Correlation

      GEREAD        0.076


COVARIANCE COVERAGE OF DATA

Minimum covariance coverage value    0.100


        PROPORTION OF DATA PRESENT

              Covariance Coverage
                  GEREAD
                  _____
GEREAD            1.000
```

```
THE MODEL ESTIMATION TERMINATED NORMALLY


MODEL FIT INFORMATION

Number of Free Parameters                          3

Loglikelihood

          H0 Value                      -24232.285
          H0 Scaling Correction Factor     3.3430
            for MLR
          H1 Value                      -24232.285
          H1 Scaling Correction Factor     3.3430
            for MLR


Information Criteria

          Akaike (AIC)                   48470.570
          Bayesian (BIC)                 48492.429
          Sample-Size Adjusted BIC       48482.895
            (n* = (n + 2) / 24)


Chi-Square Test of Model Fit

          Value                              0.000*
          Degrees of Freedom                      0
          P-Value                            0.0000
          Scaling Correction Factor          1.0000
            for MLR
```

*   The chi-square value for MLM, MLMV, MLR, ULSMV, WLSM and
    WLSMV cannot be used for chi-square difference testing in
    the regular way. MLM, MLR and WLSM chi-square difference
    testing is described on the Mplus website. MLMV, WLSMV,
    and ULSMV difference testing is done using the DIFFTEST
    option.

```
RMSEA (Root Mean Square Error Of Approximation)

          Estimate                           0.000

CFI/TLI

          CFI                                0.000
          TLI                                1.000
```

Chi-Square Test of Model Fit for the Baseline Model

      Value   0.000
      Degrees of Freedom   0
      P-Value   1.0000

SRMR (Standardized Root Mean Square Residual)

      Value for Within   0.000
      Value for Between   0.000

MODEL RESULTS

	Estimate	S.E.	Est./S.E.	Two-Tailed P-Value
Within Level				
Variances				
GEREAD	5.086	0.195	26.091	0.000
Between Level				
Means				
GEREAD	4.319	0.056	77.437	0.000
Variances				
GEREAD	0.417	0.061	6.868	0.000

QUALITY OF NUMERICAL RESULTS

      Condition Number for the Information Matrix   0.313E-01
      (ratio of smallest to largest eigenvalue)

DIAGRAM INFORMATION

  Mplus diagrams are currently not available for multilevel analysis.
No diagram output was produced.

    Beginning Time: 14:19:28
       Ending Time: 14:19:29
      Elapsed Time: 00:00:01

```
MUTHEN & MUTHEN
3463 Stoner Ave.
Los Angeles, CA  90066

Tel: (310) 391-9971
Fax: (310) 391-8971
Web: www.StatModel.com
Support: Support@StatModel.com
```

Although this is a null model in which there is not an independent variable, the output does provide some useful information that will help us understand the structure of the data. In particular, the values that are of primary interest for the null model are the Akaike's information criterion (AIC) and Bayesian information criterion (BIC) values (in the MODEL FIT INFORMATION section), which will be useful in comparing this model with others that include one or more independent variables, as we will see later. In addition, the null model also provides the estimates of the variance within the individuals (σ^2) and between the clusters (τ^2). In turn, these values can be used to estimate ρ_I (the intraclass correlation [ICC]), as in Equation 2.5. Here, the value would be

$$\widehat{\rho_I} = \frac{0.417}{0.417 + 5.086} = 0.0758$$

We interpret this value to mean that the correlation of reading test scores among students within the same schools is 0.076, if we round our result.

In order to fit the model with vocabulary as the independent variable, we would use the following syntax:

```
TITLE: 3.1 Simple Multilevel Models. Adding Level 1 Predictors
DATA: FILE IS Achieve_ch3.csv;
VARIABLE: NAMES ARE corp school class classID gender age
geread gevocab clenroll senroll;
        USEVARIABLES ARE school, geread, gevocab;
        MISSING ARE .;
        WITHIN ARE gevocab;
        CLUSTER = school;

ANALYSIS: TYPE = TWOLEVEL;

MODEL:    %WITHIN%
          geread ON gevocab;
```

Now that we are using a predictor variable, we need to define both our outcome and our predictor in the USEVARIABLES statement. Vocabulary is a level 1 variable (on the same level as our outcome). Level 1 variables are defined using the WITHIN statement. Thus, the statement WITHIN ARE gevocab defines gevocab as a level 1 or within-cluster predictor of geread. The CLUSTER statement is the same as in the null model and will remain the same for all two-level models with the structure of students nested within schools.

ANALYSIS type is still TWOLEVEL because we are still working with a basic two-level model in which the intercept is the only random effect. However, now we have a level 1 predictor (gevocab); thus, we need to define how gevocab is related to our outcome in the MODEL statement. We want to define that gevocab is predicting geread, or in other words, geread is regressed on gevocab. The MODEL statement for multilevel models in Mplus can be divided into a %WITHIN% cluster statement and a %BETWEEN% cluster statement. As mentioned before, level 1 variables are considered within-cluster variables by Mplus. Logically, then, level 2 variables are considered between-cluster variables. In this example, because we only have one predictor and it is a level 1 predictor, we will only need the %WITHIN% cluster part of the model statement. If we want to define geread being regressed on gevocab, under the %WITHIN% part of the statement we will define geread ON gevocab (for geread regressedcon gevocab). Excerpts of the output generated from this model appear below. For now we will focus on the parameter estimates and return to the interpretation of model fit and model comparison later in the chapter.

MODEL RESULTS

	Estimate	S.E.	Est./S.E.	Two-Tailed P-Value
Within Level				
GEREAD ON				
GEVOCAB	0.509	0.015	33.654	0.000
Residual Variances				
GEREAD	3.815	0.145	26.361	0.000
Between Level				
Means				
GEREAD	2.043	0.067	30.711	0.000
Variances				
GEREAD	0.105	0.017	6.057	0.000

In interpreting the output, it is helpful to consider what effects we expect to see. The equation for the full model in this analysis would be written as

$$\text{geread} = \gamma_{00} + \gamma_{10}(\text{gevocab}) + u_{0j} + \varepsilon_{ij}$$

Thus, we would expect to find the values for the two fixed effects (γ_{00} and γ_{10}) as well as for the two error terms (u_{0j} and ε_{ij}). Starting with γ_{00}, recall that γ_{00} is the intercept of the intercept, or the overall intercept of the prediction model. On the Mplus output, γ_{00} can be found on the Between Level as the estimate labeled Means GEREAD. Here, the intercept of the prediction model is 2.043, and it is significant ($p < .001$), indicating that the intercept is significantly different from zero. Generally of more interest are the slope parameter estimates. In particular, in this model, γ_{10} is the slope for vocabulary. The slope for gevocab can be found on the Within Level as the estimate labeled GEREAD ON GEVOCAB. The slope for gevocab is 0.509 and is statistically significant ($p > .001$), indicating that it is a significant predictor of geread, and that as vocabulary score increases by 1 point, reading ability increases by 0.509 points.

In addition to the fixed effects in model 3.1, we can also ascertain how much variation in geread is present across schools. The random effect for the intercept (u_{0j}) can be found on the Between Level as the estimate labeled Variances. The output shows that after accounting for the impact of gevocab, the estimate of variation in intercepts across schools is 0.105. The variation in the intercepts is also statistically significant ($p < .001$), indicating that there is significant variation in intercepts across schools. The within-school variation (r_{ij}), found on the Within Level as the estimate labeled Residual Variances, is estimated as 3.815. Using these variance estimates, we can tie these numbers directly back to our discussion in Chapter 2, where $\tau_0^2 = 0.105$ and $s^2 = 3.815$. In addition, the overall fixed intercept, denoted as γ_{00} in Chapter 2, is 2.043, which is the mean of geread when the gevocab score is 0.

Let us now consider a similar model with a single predictor. This time, however, our single predictor will be a school-level characteristic (level 2 predictor) instead of an individual-level characteristic (level 1 predictor). Definition of such a model is extremely similar to that of our previous model, except that our level 2 predictor is defined in the %BETWEEN% statement rather than the %WITHIN% statement.

```
TITLE: 3.2 Simple Multilevel Models. Adding Level 1 and
Level 2 Predictors
DATA: FILE IS Achieve_ch3.csv;
VARIABLE: NAMES ARE corp school class classID gender age
geread gevocab clenroll senroll;
          USEVARIABLES ARE school geread senroll;
          MISSING ARE .;
          BETWEEN ARE senroll;
          CLUSTER = school;

ANALYSIS: TYPE = TWOLEVEL;

MODEL:    %BETWEEN%
          geread ON senroll;
```

Model results output for this model appears below.

MODEL RESULTS

	Estimate	S.E.	Est./S.E.	Two-Tailed P-Value
Within Level				
Variances				
GEREAD	5.086	0.195	26.091	0.000
Between Level				
GEREAD ON				
SENROLL	0.000	0.000	0.182	0.856
Intercepts				
GEREAD	4.292	0.158	27.203	0.000
Residual Variances				
GEREAD	0.417	0.061	6.877	0.000

Using a single level 2 predictor, the output looks very much like the previous example (with a single level 1 predictor), except now the slope coefficient for our level 2 predictor, senroll, can be found on the Between Level under GEREAD ON SENROLL. The slope for GEREAD (γ_{01}) is not significant ($p = .856$), indicating that school enrolment is not a significant predictor of reading ability. Like the previous model, there is a significant overall intercept ($\gamma_{00} = 4.292$, $p < .001$), indicating that the intercept is significantly different from zero and a significant random effect for the intercept ($u_{0j} = 5.086$, $p < .001$), indicating significant variation in the intercept across schools.

The models in the previous examples were quite simple, only incorporating one level 1 predictor. In most applications, however, researchers will have multiple predictor variables, often at both level 1 and level 2. Incorporation of multiple predictors at different levels of analysis in Mplus is done in much the same manner as incorporation of a single predictor. For example, let us assume that the researcher now wants to use both the size of the school (senroll) and the vocabulary (gevocab) to predict the overall reading score. In that instance, the Mplus syntax would result:

```
TITLE: 3.3 Simple Multilevel Models. Adding level 1 and
level 2 predictors
DATA: FILE IS Achieve_ch3.csv;
VARIABLE: NAMES ARE corp school class classID gender age
geread gevocab clenroll senroll;
        USEVARIABLES ARE school geread gevocab senroll;
        MISSING ARE .;
        WITHIN ARE gevocab;
        BETWEEN ARE senroll;
        CLUSTER = school;
```

```
ANALYSIS: TYPE = TWOLEVEL;

MODEL:      %WITHIN%
            geread ON gevocab;

            %BETWEEN%
            geread on senroll;
```

And the following output would result:

```
MODEL RESULTS
```

	Estimate	S.E.	Est./S.E.	Two-Tailed P-Value
Within Level				
GEREAD ON				
GEVOCAB	0.509	0.015	33.672	0.000
Residual Variances				
GEREAD	3.815	0.145	26.364	0.000
Between Level				
GEREAD ON				
SENROLL	0.000	0.000	-0.137	0.891
Intercepts				
GEREAD	2.055	0.115	17.893	0.000
Residual Variances				
GEREAD	0.105	0.017	6.006	0.000

From these results, we can see when both vocabulary and school enrolment are used to predict reading, vocabulary had a statistically significant relationship with reading ($\gamma_{10} = .509$, $p < .001$), whereas enrolment did not have a statistically significant relationship with reading achievement ($\gamma_{01} = .000$, $p < .891$). In addition, notice that there were some minor changes in the estimates of the other model parameters from the other two models, which are to be expected because both predictor variables are now being used simultaneously.

Random Coefficients Models

In Chapter 2, we described the random coefficients model, in which the impact of the independent variable on the dependent is allowed to vary across the level 2 effects. In the context of this research problem, this would mean that we could allow the impact of gevocab on geread to vary from one school to another. Incorporating such random coefficients effects into a multilevel model using Mplus requires a different analysis type. To define a random coefficients model, in the ANALYSIS line of the input the TYPE will now be TWOLEVEL RANDOM instead of TWOLEVEL. Let us return to the scenario in model 3.1 but this time allow both the slope and the intercept for gevocab to vary randomly from one school to another. In order to define a random slope for gevocab, the definition of the main effect for gevocab will change. It will no longer be defined on the %WITHIN% line. Instead, on the %WITHIN% line we will create a variable to represent the random slope. In this example, we name the random coefficients effect for gevocab gevocab_sl (though you can name it whatever will make the most sense to you). The line gevocab_sl | geread on gevocab defines gevocab_sl as the random slope of geread regressed on gevocab.

Although we have in essence defined gevocab as a predictor of geread, and allowed it to vary randomly, there is one more nuance to this model. When a model only contains a random intercept, we only need to be concerned about the random intercept (τ_{00}). When random effects are incorporated into a multilevel model, the variances and covariances of these random effects are described by the tau matrix. In this example where we have a random intercept and one random slope, the tau matrix would look like

$$\tau = \begin{bmatrix} \tau_{00} & \tau_{10} \\ \tau_{01} & \tau_{11} \end{bmatrix} \tag{3.1}$$

where:
  τ_{00} is the variance of the random intercept
  τ_{11} is the variance of the random slope
  τ_{01}/τ_{10} is the covariance of the random intercept and random slope

In Mplus, this matrix is not actually allowed to vary freely by default. Instead, the Mplus default is for the tau matrix to be a diagonal matrix, or in other words, the default assumption is that the covariance between the random intercept and the random slope is zero.

$$\tau = \begin{bmatrix} \tau_{00} & 0 \\ 0 & \tau_{11} \end{bmatrix} \tag{3.2}$$

It is possible, however, to override this default by specifying the correlation between the random intercept and the random slope in the model syntax. On the %BETWEEN% level, we will define this relationship with the syntax geread with gevocab_sl. The syntax for this model would now becomes

```
TITLE: 3.5 Simple Multilevel Models. Random Coefficients Model
DATA: FILE IS Achieve_ch3.csv;
VARIABLE: NAMES ARE corp school class classID gender age
geread gevocab clenroll senroll;
        USEVARIABLES ARE school geread gevocab;
```

```
        MISSING ARE .;
        WITHIN ARE gevocab;
        CLUSTER = school;

ANALYSIS: TYPE = TWOLEVEL RANDOM;

MODEL:    %WITHIN%
          gevocab_sl | geread on gevocab;


          %BETWEEN%
          geread with gevocab_sl;
```

Output for this model appears below. It should be noted, however, that sometimes the researcher may not wish to estimate the covariance between the random intercept and the random slope (τ_{10}). If this is the case, this with statement may simply be left out and Mplus will simply estimate this relationship to be zero.

MODEL RESULTS

	Estimate	S.E.	Est./S.E.	Two-Tailed P-Value
Within Level				
Residual Variances				
GEREAD	3.716	0.140	26.516	0.000
Between Level				
GEREAD WITH				
GEVOCAB_SL	-0.065	0.013	-4.968	0.000
Means				
GEREAD	2.024	0.063	32.173	0.000
GEVOCAB_SL	0.518	0.014	36.334	0.000
Variances				
GEREAD	0.315	0.067	4.720	0.000
GEVOCAB_SL	0.019	0.004	5.237	0.000

Now examining the results of the analysis, we find the effect for gevocab on geread on the Between Level under Means GEVOCAB_SL. From the output, we can see that gevocab is statistically significantly related to geread across schools. The estimated coefficient, 0.518, corresponds to γ_{10} from Chapter 2 and is interpreted as the average impact of the predictor on the outcome across schools. In this model, we designated the slope for gevocab to be random. Looking under Variances GEVOCAB_SL, we can see that there is significant variation in the slopes for gevocab among the schools in our sample. This estimate indicates that the coefficient for gevocab varies from one school to another; that is, the relationship of the independent and dependent variables differs across schools. We can also see the relationship between the slope for gevocab and the intercept. On the Between Level under GEREAD with GEVOCAB_SL, we find the value for τ_{11}, indicating that there is a statistically significant negative relationship between the intercept and the slope for vocabulary. In other words, schools for which the relationship between the variables was stronger had lower means on the dependent variable.

A model with two random slopes can be defined in much the same way as for a single slope. As an example, suppose that our researcher is interested in determining whether the gender of the student (with 0 coded female and 1 coded male) also impacted reading performance and wants to allow this effect to vary from one school to another. Syntax for this model appears below. Here we see that vocabulary is significantly related to reading ($\gamma_{01} = .518, p < .001$), whereas gender is not significantly related to geread ($\gamma_{02} = -.014, p = .715$). In addition, the random variance of coefficients for vocabulary across schools is significant ($\tau_{10} = .019, p < .001$), indicating that the impact of vocabulary on reading ability varies significantly from school to school.

```
TITLE: 3.5 Simple Multilevel Models. Two Random Slopes
DATA: FILE IS Achieve_ch3.csv;
VARIABLE: NAMES ARE corp school class classID gender age
geread gevocab clenroll senroll;
        USEVARIABLES ARE school gender geread gevocab;
        MISSING ARE .;
        WITHIN ARE gender gevocab;
        CLUSTER = school;

ANALYSIS: TYPE = TWOLEVEL RANDOM;

MODEL:    %WITHIN%
        gevocab_sl | geread on gevocab;
        gender_sl | geread on gender;

        %BETWEEN%
        geread with gevocab_sl;
        geread with gender_sl;
        gevocab_sl with gender_sl;
```

MODEL RESULTS

	Estimate	S.E.	Est./S.E.	Two-Tailed P-Value
Within Level				
Residual Variances				
GEREAD	3.712	0.140	26.441	0.000

```
Between Level
  GEREAD    WITH
     GEVOCAB_SL        -0.062        0.013       -4.844        0.000
     GENDER_SL          0.038        0.038        0.986        0.324

GEVOCAB_WITH
     GENDER_SL         -0.005        0.007       -0.832        0.405

Means
     GEREAD             2.015        0.063       31.883        0.000
     GEVOCAB_SL         0.518        0.014       36.453        0.000
     GENDER_SL          0.014        0.037        0.366        0.715

Variances
     GEREAD             0.268        0.061        4.419        0.000
     GEVOCAB_SL         0.019        0.004        5.202        0.000
     GENDER_SL          0.019        0.035        0.527        0.598
```

Interactions and Cross-Level Interactions

Interactions among the predictor variables, and in particular cross-level interactions, can be very important features in the application of multi-level models. Cross-level interactions occur when the impact of a level 1 variable on the outcome (e.g., vocabulary score) differs depending on the value of the level 2 predictor (e.g., school enrollment). Interactions, be they within the same level or cross-level, are simply the product of two predictors. Thus, incorporation of interactions on the same level of analysis is done in exactly the same manner as would be used for a multiple regression model. A new variable, which is the product of the two interacting variables is created and used as a predictor in the analysis. Considering interactions between predictors on differing levels of analysis, however, is a slightly different process in Mplus and will look different depending on whether you want to work with a random coefficients or a random intercept model.

Both models 3.5 and 3.6 are models defining cross-level interactions in Mplus. Model 3.5 defines a multilevel model in which level 1 predictor vocabulary is interacting with level 2 predictor school enrolment in which vocabulary is allowed to vary randomly (a random coefficients model). Model 3.6 defines the same cross-level interaction in which level 1 predictor vocabulary is interacting with level 2 predictor school enrolment, but in this model there is only a random intercept.

```
TITLE: 3.6a Simple Multilevel Models. Cross Level Interactions
with Random Coefficients
DATA: FILE IS Achieve_ch3.csv;
VARIABLE: NAMES ARE corp school class classID gender age
geread gevocab clenroll senroll;
        USEVARIABLES ARE school geread gevocab senroll;
        MISSING ARE .;
        WITHIN ARE gevocab;
        BETWEEN ARE senroll;
        CLUSTER = school;

ANALYSIS: TYPE = TWOLEVEL RANDOM;

MODEL:    %WITHIN%
          gevocab_sl | geread on gevocab;

          %BETWEEN%
          geread gevocab_sl on senroll;

TITLE: 3.6b Simple Multilevel Models. Cross Level Interactions
with Random Intercepts
DATA: FILE IS Achieve_ch3.csv;
VARIABLE: NAMES ARE corp school class classID gender age
geread gevocab clenroll senroll;
        USEVARIABLES ARE school geread gevocab senroll;
        MISSING ARE .;
        WITHIN ARE gevocab;
        BETWEEN ARE senroll;
        CLUSTER = school;

ANALYSIS: TYPE = TWOLEVEL RANDOM;

MODEL:    %WITHIN%
          gevocab_sl | geread on gevocab;

          %BETWEEN%
          geread gevocab_sl on senroll;
          gevocab_sl @ 0;
```

As can be seen, regardless of whether random coefficients or random intercepts are desired, defining models with cross-level interactions requires running the model as TWOLEVEL RANDOM. In other words, in order to define cross-level interactions, first the model must be specified as a random coefficients model. Then if only random intercepts are desired, additional commands to modify the model are required. Specifically, a statement is necessary to set the variance of the random slope to zero. In order to do this, the following syntax may be used: gevocab_sl @ 0. Output for both models 3.5 and 3.6 appear as follows.

Model 3.6a Results

MODEL RESULTS

	Estimate	S.E.	Est./S.E.	Two-Tailed P-Value
Within Level				
Residual Variances				
GEREAD	3.759	0.141	26.591	0.000
Between Level				
GEVOCAB_SL ON				
SENROLL	0.000	0.000	-0.885	0.376
GEREAD ON				
SENROLL	0.000	0.000	0.967	0.334
Intercepts				
GEREAD	1.829	0.216	8.450	0.000
GEVOCAB_SL	0.553	0.052	10.719	0.000
Residual Variances				
GEREAD	0.052	0.025	2.102	0.036
GEVOCAB_SL	0.005	0.001	4.471	0.000

Looking at the output from both models 3.5 and 3.6, the cross-level interaction appears on the Between level labeled GEVOCAB_SL ON SENROLL. Focusing on model 3.6a, we can see that vocabulary is a significant predictor of reading ($\gamma_{01} = .553$, $p < .001$); however, neither school enrolment ($\gamma_{10} = .000$, $p = .333$) nor the interaction between vocabulary and school enrolment ($\gamma_{11} = .000$, $p = .376$) was significant. Looking at the random effects, both the variances of the random intercept ($\tau_{00} = .052$, $p < .036$) and the random slope ($\tau_{10} = .005$, $p < .001$) were significant, indicating that the intercept was varied significantly across schools, and the impact of vocabulary on reading varied significantly across schools.

Model 3.6b Results

MODEL RESULTS

	Estimate	S.E.	Est./S.E.	Two-Tailed P-Value
Within Level				
Residual Variances				
GEREAD	3.810	0.144	26.383	0.000

```
Between Level

GEVOCAB_SL  ON
    SENROLL            0.000      0.000     -1.274       0.203

GEREAD      ON
    SENROLL            0.001      0.000      1.267       0.205

Intercepts

    GEREAD             1.751      0.216      8.089       0.000
    GEVOCAB_SL         0.576      0.050     11.429       0.000

Residual Variances

    GEREAD             0.101      0.018      5.730       0.000
    GEVOCAB_SL         0.000      0.000    999.000     999.000
```

The output from model 3.6 has a similar interpretation. Looking at the fixed effects, we see the same pattern as model 3.5: Vocabulary is a significant predictor of reading; however, school enrolment and the interaction of vocabulary and school enrolment were not significant predictors of reading. We see a difference, however, when we look to the model random effects. There is still significant variance in the random intercept ($\tau_{00} = .101, p < .001$); however, where the random slope is listed, we see the variance listed as 0.000 and the p-value at 999. This is because in this model we wanted it run as random intercepts only, so the variance of the random slope was set to zero prior to running the model. This chapter provided a basic overview to running random intercepts and random coefficients models using Mplus. In Chapter 4, the issues of model fit, model comparison, and customizable options will be considered in some detail.

Summary

In this chapter, we learned how to apply the concepts that were introduced in Chapter 2, using the Mplus software environment. We progressed from the simple random intercepts model to analyses that accommodate random coefficients and predictors at both levels 1 and 2. In Chapter 4, we extend these ideas by exploring the issue of centering and by learning more about how to control output from Mplus so that it is tailored to address the questions that are of primary interest in a given research scenario. In Chapter 5, we extend the two-level model to accommodate three levels.

4

Additional Issues in Fitting Two-Level Models

Model Comparison and Model Fit

In Chapter 3, we discussed defining simple multilevel models using Mplus syntax and interpreting the multilevel model parameters provided in the Mplus output. In our discussion thus far, however, we have only focused on the values and interpretations of the model parameters (fixed and random effects) but have given no thought yet to the overall predictive value of the model. Looking back at the Mplus output provides us with some useful information regarding the structure and fit of the model. In particular, the values that are of primary interest can be found in the MODEL FIT INFORMATION section. This section provides statistics that will be useful in comparing the fit of a current model with other models.

When discussing the predictive value, or fit of a multilevel model, the concept of R squared (as discussed previously in Chapter 2) is often applied as a measure of the fit of a single model. However, more often than not, model fit is not discussed as just the fit of a single model, but rather the comparative fit of a model versus other competing models. This is the idea of model comparison. Often, a researcher is not sure of the optimal predictor set or structure for a multilevel model so instead of just fitting one model, a systematic series of models is fit and compared in order to find the best comparative fit. Models being compared can be nested or non-nested in nature. Nested models involve a series of systematic models in which each less complex model is contained in each model of higher complexity. A very common way to build nested models is to begin with a null model (see Chapter 3 for definition) to get a baseline for model fit when no predictors are used, and then to add predictors and complex terms systematically in order to see if the addition of parameters significantly improves model fit compared with the null model. Non-nested models do not need to be as systematic and can contain models with differing predictor sets and structures. Depending on the type of comparison (nested or non-nested), different model comparison statistics are relevant.

Model Comparison for Non-Nested Models

The simplest and most general way to do model comparisons for multilevel models involves statistics for non-nested models. The most common statistics for comparing non-nested models are the Akaike information criterion (AIC) and the Bayesian information criterion (BIC). The AIC and BIC are general measures of the fit of model which correct for the complexity of the model. A smaller value for AIC/BIC indicates a better fit model. It should be noted that the AIC and BIC are only interpretable in comparison with other AIC and BIC statistics. They have no inherent meaning when taken in isolation. AIC and BIC also have no associated significance test or effect size; thus, a statement can only be made about which the model fits better, but it cannot be determined how much better or if the fit is significantly better. Let us return to an example from Chapter 3 (model 3.1) where we predicted reading ability from vocabulary. We previously focused our discussion on the Model Results section where we talked about the interpretation of model parameters. Now let us focus on the Model Fit Information section (located several sections above Model Results section).

```
Model 3.1
MODEL FIT INFORMATION

Number of Free Parameters                         4

Loglikelihood

        H0 Value                          -22615.828
        H0 Scaling Correction Factor          3.2492
          for MLR
        H1 Value                          -22615.828
        H1 Scaling Correction Factor          3.2492
          for MLR

Information Criteria

        Akaike (AIC)                       45239.655
        Bayesian (BIC)                     45268.801
        Sample-Size Adjusted BIC           45256.089
          (n* = (n + 2) / 24)

Chi-Square Test of Model Fit

        Value                                 0.000*
        Degrees of Freedom                         0
        P-Value                               0.0000
        Scaling Correction Factor             1.0000
          for MLR
```

* The chi-square value for MLM, MLMV, MLR, ULSMV, WLSM and WLSMV cannot be used for chi-square difference testing in the regular way. MLM, MLR and WLSM chi-square difference testing is described on the Mplus website. MLMV, WLSMV, and ULSMV difference testing is done using the DIFFTEST option.

RMSEA (Root Mean Square Error Of Approximation)

            Estimate   0.000

CFI/TLI

            CFI   1.000
            TLI   1.000

Chi-Square Test of Model Fit for the Baseline Model

            Value   1089.645
            Degrees of Freedom   1
            P-Value   0.0000

SRMR (Standardized Root Mean Square Residual)

            Value for Within   0.000
            Value for Between   0.000

As shown in the Model Fit Information for model 3.1, according to the Information Criteria section, this model has AIC = 45239 and BIC = 45268. As previously mentioned, these values carry no model fit information if interpreted alone; however, if we compare these values to the AIC and BIC of another model, we can determine which model provides the better fit. Let us compare this, for example, with a similar model predicting vocabulary, but instead of using reading ability as our predictor (Model 4.1), we want to see if a model using age as the predictor provides a better fit.

```
TITLE: 4.1
DATA: FILE IS Achieve_ch3.csv;
VARIABLE: NAMES ARE corp school class classID gender age
geread gevocab clenroll senroll;
          USEVARIABLES ARE school, geread, age;
          MISSING ARE .;
          WITHIN ARE age;
          CLUSTER = school;
```

```
ANALYSIS: TYPE = TWOLEVEL;

MODEL:    %WITHIN%
           geread ON age;
```

Information Criteria

Akaike (AIC)	48448.878
Bayesian (BIC)	48478.024
Sample-Size Adjusted BIC	48465.312
(n* = (n + 2) / 24)	

MODEL RESULTS

	Estimate	S.E.	Est./S.E.	Two-Tailed P-Value
Within Level				
GEREAD ON				
AGE	-0.021	0.004	-5.158	0.000
Residual Variances				
GEREAD	5.077	0.195	26.053	0.000
Between Level				
Means				
GEREAD	6.612	0.448	14.755	0.000
Variances				
GEREAD	0.407	0.059	6.935	0.000

Model 4.1 yields AIC = 48448 and BIC = 48478. According to the AIC and BIC, model 3.1 has lower AIC and BIC compared to model 4.1 so model 3.1 is the better fit.

Model Comparison for Nested Models

When comparing nested models, the most common method of model comparison is the deviance test. Similar to AIC and BIC, deviance is a quantification of the fit of a model. When models are nested, a chi-square significance test can be used to compare the deviances of the nested models to determine if one model fits significantly better than the other. Unfortunately, Mplus does not include a chi-square deviance test, which is appropriate for multilevel models on its printout, nor does it provide the deviance values. As mentioned in Chapter 2, Mplus uses maximum likelihood with robust standard errors (MLR) to estimate multilevel models. Although not inherently a problem, the use of MLR estimation means that the computation of the log likelihood and chi-square tests of model fit presented on the Mplus printout is not accurate for multilevel models. A note on the printout below the chi-square test calls attention to this issue.

```
*    The chi-square value for MLM, MLMV, MLR, ULSMV, WLSM and
     WLSMV cannot be used for chi-square difference testing in
     the regular way. MLM, MLR and WLSM chi-square difference
     testing is described on the Mplus website. MLMV, WLSMV, and
     ULSMV difference testing is done using the DIFFTEST option.
```

As indicated in the note, a procedure for how to obtain a chi-square difference test is described on the Mplus website. The procedure on the website describes how to compute the chi-square difference test for MLR models based on the information provided on the Mplus printout. We will briefly describe this procedure here as well.

1. The first step is to run the desired model(s) in Mplus. Once an output is obtained, the relevant information will be the log likelihood values, scaling correction factors and parameters for the two comparison models, (H0 and H1). If the comparison is with the null model, the values for both H0 and H1 can be found on the printout for H1.

2. Once relevant information is obtained from the printout(s), the difference test scaling correction can be computed as

$$cd = \frac{\left[(p0)*(c0)-(p1)*(c1)\right]}{(p0-p1)} \tag{4.1}$$

where:
   $p0$ is the number of models for the nested model
   $p1$ is the number of models for the comparison model

3. Once the difference test scaling correction has been computed, the adjusted chi-square difference test (TRd) can be computed as

$$TRd = \frac{-2*(L0-L1)}{cd} \tag{4.2}$$

The resulting statistic is distributed as a chi-square with the $p1 - p0$ degrees of freedom.

The necessary information to use the formulas presented in the Model Fit for Nested Models section can be obtained from the printout in the Model Fit Information section under the header Loglikelihood.

Model 3.2 MODEL FIT INFORMATION

MODEL FIT INFORMATION

Number of Free Parameters   5

Loglikelihood

                H0 Value   -22615.820
                H0 Scaling Correction Factor   2.7615
                   for MLR
                H1 Value   -22615.820
                H1 Scaling Correction Factor   2.7615
                   for MLR

Information Criteria

                Akaike (AIC)   45241.640
                Bayesian (BIC)   45278.072
                Sample-Size Adjusted BIC   45262.183
                   (n* = (n + 2) / 24)

For a complete application example, see the companion website (https://www.crcpress.com/Multilevel-Modeling-Using-Mplus/Finch-Bolin/p/book/9781498748247).

Centering Predictors

As per the discussion in Chapter 2, it may be advantageous to center predictors, especially when interactions are incorporated. Centering predictors can both provide slightly easier interpretation of interaction terms as well as help alleviate issues of multicollinearity arising from inclusion of both main effects and interactions in the same model. Recall that centering of a variable entails the subtraction of a mean value from each score on the variable. In earlier versions of Mplus, centering of predictors can be accomplished through use of the Centering command (i.e., CENTERING = GRANDMEAN(gevocab);). In version 7 of Mplus, the syntax for centering changes to the Define command (i.e., DEFINE: CENTER geread(GRANDMEAN);). For an example, let us use a very simple model predicting geread from gevocab. In 4.2a, we can see a completely uncentered version of this model. In 4.2b, we can see a version of this model in which the outcome variable (gevocab) has been grand mean centered.

```
TITLE: 4.2a Centering Predictors: Uncentered Model
DATA: FILE •IS Achieve_ch3.csv;
VARIABLE: NAMES ARE corp school class classID gender age
geread gevocab clenroll senroll;
        USEVARIABLES ARE school, geread, gevocab;
        MISSING ARE .;
        WITHIN ARE gevocab;
        CLUSTER = school;

ANALYSIS: TYPE = TWOLEVEL;

MODEL:    %WITHIN%
           geread ON gevocab;
```

Information Criteria

Akaike (AIC)	45239.655
Bayesian (BIC)	45268.801
Sample-Size Adjusted BIC	45256.089
(n* = (n + 2) / 24)	

MODEL RESULTS

	Estimate	S.E.	Est./S.E.	Two-Tailed P-Value
Within Level				
GEREAD ON				
GEVOCAB	0.509	0.015	33.654	0.000
Residual Variances				
GEREAD	3.815	0.145	26.361	0.000
Between Level				
Means				
GEREAD	2.043	0.067	30.711	0.000
Variances				
GEREAD	0.105	0.017	6.057	0.000

```
TITLE:  4.2b Centering Predictors
DATA: FILE IS Achieve_ch3.csv;
VARIABLE: NAMES ARE corp school class classID gender age
geread gevocab clenroll senroll;
          USEVARIABLES ARE school, geread, gevocab;
          MISSING ARE .;
          WITHIN ARE gevocab;
          CLUSTER = school;
DEFINE:   CENTER geread(grandmean);

ANALYSIS: TYPE = TWOLEVEL;

MODEL:    %WITHIN%
          geread ON gevocab;
```

Information Criteria

Akaike (AIC)	45239.655
Bayesian (BIC)	45268.801
Sample-Size Adjusted BIC	45256.089
(n* = (n + 2) / 24)	

MODEL RESULTS

	Estimate	S.E.	Est./S.E.	Two-Tailed P-Value
Within Level				
GEREAD ON				
GEVOCAB	0.509	0.015	33.654	0.000
Residual Variances				
GEREAD	3.815	0.145	26.361	0.000
Between Level				
Means				
GEREAD	-2.317	0.067	-34.831	0.000
Variances				
GEREAD	0.105	0.017	6.057	0.000

First, notice the identical model fit (compare AIC/BIC and log likelihood) of the centered and uncentered models. This is a good way to check to be sure the centering worked. Now, looking to the fixed effects of the model, we see that the only difference in these models is in the intercept. This is to be expected because the grand mean centering has changed the scale of this variable. However, only the scale has been changed; thus, the *t*-value (Estimate/Standard Error [Est/S.E.]) and *p*-value are identical to the original model, again demonstrating the equivalence of the models.

Summary

This chapter has demonstrated how Mplus can be used to compare model fit and how variables can be centered. In Chapter 5, we will see how to use Mplus to run multilevel models for longitudinal data and expand our two-level models into three-level models. Many of the ideas described here will also be applicable in those cases. Indeed, even as we expand our discussion to models in which the dependent variables are not continuous, and to situations in which the variables of interest are not observed, but rather latent, the tools that we have worked on together in this chapter will remain quite useful.

5

Fitting Three-Level Models in Mplus

In Chapters 2 and 3, we introduced the multilevel modeling framework and demonstrated the use of the Mplus in fitting two-level models. In Chapter 5, we will expand upon this basic two-level framework by fitting models with additional levels of data structure. As described in Chapter 2, it is conceivable for a level 1 unit, such as students, to be nested in higher level units, such as classrooms. Thus, in keeping with our examples, we might assume that at least a portion of a students' performance on a reading test is due to the classroom in which they find themselves. Each classroom may have a unique learning context, which might contribute to the performance of the students in the classroom, such as the quality of the teacher, the presence of disruptive students, and the time of day in which students are in the class, among others. Furthermore, as we saw in Chapters 2 and 3, the impact of fixed effects on the dependent variable can vary among the level 2 units, resulting in a random slopes model. Now we will see that, using Mplus, it is possible to estimate models with three levels of a nested structure and that the syntax for defining and fitting these models is very similar to that which we used in the two-level case.

In this chapter, we will continue working with the data described in Chapter 3. In our Chapter 3 examples, we included two levels of data structure, students within schools, along with associated predictors of reading achievement at each level. We will now add a third level of structure, classroom, which is nested within schools. In this context, nested simply means that students within a given classroom will all be found within the same school. Thus, the data are such that students are nested within classrooms, which are in turn nested within schools. Before we begin defining the syntax for three-level models, however, it is important to call attention to how ID numbers are applied for the different levels of analysis. When using Mplus for multilevel models, it is important that all ID numbers be unique. This has not been an issue before because we just had two levels of analysis. However, when you add a third level of analysis, it is imperative that the middle level (level 2) have completely unique IDs. For example, we have 100 schools each with 15 classrooms a piece. If each classroom is numbered 1–15 within each school, Mplus would have a problem because the classroom IDs are not completely unique. You would want to make sure to recode the classroom IDs before moving forward.

Defining Simple Three-Level Models Using Mplus Syntax

The Mplus syntax for defining and fitting models incorporating three levels of data structure is very similar to that for two-level models that we have already seen. Let us begin by defining a null model for prediction of student reading achievement, where regressors might include student-level characteristics, classroom-level characteristics, and school-level characteristics. The syntax to fit a three-level null model appears is as follows:

```
TITLE: 5.1 Three Level Null level
DATA: FILE IS Achieve_ch3.csv;
VARIABLE: NAMES ARE corp school class classID gender age
geread gevocab clenroll senroll;
        USEVARIABLES ARE school, classID, geread;
        MISSING ARE .;
        CLUSTER = school classID;

ANALYSIS: TYPE = THREELEVEL;

MODEL:     %WITHIN%
            geread;
```

As can be seen, the syntax for fitting a random intercept model with three levels is very similar to that for the random intercept model with two levels, except we will now use the TYPE = THREELEVEL Analysis type, and when defining our clusters (levels) in the VARIABLE command, we will have to define clusters for both school and classroom (class) levels.

```
Information Criteria

            Akaike (AIC)                    48350.443
            Bayesian (BIC)                  48379.589
            Sample-Size Adjusted BIC        48366.877
              (n* = (n + 2) / 24)

MODEL RESULTS

                                                        Two-Tailed
                    Estimate      S.E.    Est./S.E.      P-Value

Within Level

    Variances
        GEREAD          4.893     0.178     27.564         0.000
```

```
Between CLASSID Level

  Variances
    GEREAD              0.266        0.077      3.446       0.001

Between SCHOOL Level

  Means
    GEREAD              4.321        0.056     77.279       0.000

  Variances
    GEREAD              0.341        0.062      5.505       0.000
```

As this is a random intercept only model, there is not much to be interpreted other than model fit (AIC, BIC). There are, however, some pieces of information to take note of. For example, we see two different sets of random effects: random effects for the between-class level ($\tau = 0.266$, $p < .001$) so that the intercept is modeled to vary across classrooms and random effects for the between-school level ($\tau = 0.341$, $p < .001$) so that the intercept is modeled to vary across schools. Remember from our discussion in Chapter 2 that we can also interpret these random intercepts as means of the dependent variable (reading) varying across levels of the random effects (classrooms and schools).

We can now add predictors to the fixed portion of a multilevel model with three or more levels in much the same manner as for a two-level model. For example, we may wish to extend the intercept only model described previously so that it includes several independent variables, such as student's vocabulary test score (gevocab), the size of the student's reading classroom (clenroll), and the size of the student's school (senroll).

```
TITLE: 5.2 Three Level Model with Predictors at each level
DATA: FILE IS Achieve_ch3.csv;
VARIABLE: NAMES ARE corp school class classID gender age
geread gevocab clenroll senroll;
            USEVARIABLES ARE school, classID, geread, gevocab,
            clenroll, senroll;
            MISSING ARE .;
            WITHIN ARE gevocab;
            BETWEEN ARE (classID) clenroll (school) senroll;
            CLUSTER = school classID;

ANALYSIS: TYPE = THREELEVEL;

MODEL:      %BETWEEN school%
                geread on senroll;

            %BETWEEN classID%
             geread on clenroll;
```

```
%WITHIN%
    geread on gevocab;
```

Information Criteria

                    Akaike (AIC)   45202.545
                    Bayesian (BIC)   45253.549
                    Sample-Size Adjusted BIC   45231.304
                        (n* = (n + 2) / 24)

MODEL RESULTS

		Estimate	S.E.	Est./S.E.	Two-Tailed P-Value
Within Level					
GEREAD	ON				
GEVOCAB		0.504	0.014	35.454	0.000
Residual Variances					
GEREAD		3.742	0.141	26.566	0.000
Between CLASSID Level					
GEREAD	ON				
CLENROLL		0.023	0.009	2.539	0.011
Residual Variances					
GEREAD		0.096	0.027	3.612	0.000
Between SCHOOL Level					
GEREAD	ON				
SENROLL		0.000	0.000	−0.542	0.588
Intercepts					
GEREAD		1.631	0.212	7.687	0.000
Residual Variances					
GEREAD		0.079	0.017	4.647	0.000

When interpreting the output, we will first want to ascertain whether including the predictor variables resulted in a better fitting model. As we saw in Chapter 4, we can compare model by examining the AIC and BIC values for each, where lower values indicate better fit. For the original null model, these values were 48350 and 48379, respectively, which are both larger than the AIC and BIC for model 5.2 (45202 and 45253). Therefore, we would conclude that this latter model, including a single predictor variable at each level, provides better fit to the data, and thus is preferable to the null model with no predictors.

We can see from the output for model 5.2 that student's vocabulary score ($t = 35.45$, $p < .001$) and size of the classroom ($t = 2.53$, $p < .05$) are statistically significant positive predictors of student reading achievement score, but the size of the school ($t = -0.542$, $p = .588$) does not significantly predict reading achievement. As a side note, the significant positive relationship between the size of the classroom and the reading achievement might seem to be a bit confusing, suggesting that students in larger classrooms had higher reading achievement test scores. However, in this particular case, larger classrooms very frequently included multiple teachers' aides, so that the actual adult-to-student ratio might be lower than in classrooms with fewer students. In addition, estimates for the random intercepts of classroom nested in school and school have decreased in value from those of the null model, and are each significant, suggesting that when we account for the three fixed effects, some of the mean differences between schools and between classrooms is accounted for.

Random Coefficients Models with Three Levels in Mplus

In Chapter 2, we discussed the random coefficients multilevel model, in which the impact of one or more fixed effects is allowed to vary across the levels of a random effect. Thus, for example, we could assess whether the relationship of vocabulary test score on reading achievement differs by school. In Chapter 3, we learned how to fit such random coefficient models using ANALYSIS TYPE = TWOLEVEL RANDOM. Given the relative similarity in syntax for fitting two-level and three-level models, as might be expected the definition of random coefficients models in the three-level context with Mplus is very much like that for two-level models. As an example, let us consider a model in which we are interested in determining whether mean reading scores differ between males and females, while accounting for the relationship between vocabulary and reading. Furthermore, we believe that the relationship of gender to reading may differ across schools and across classrooms, thus leading to a model where the gender coefficient is allowed to vary across both random effects in our three-level model. The syntax for fitting this model is as follows:

```
TITLE: 5.3 Three Level Random Coefficients Model
DATA: FILE IS Achieve_ch3.csv;
VARIABLE: NAMES ARE corp school class classID gender age
geread gevocab clenroll senroll;
```

```
        USEVARIABLES ARE school, classID, geread, gevocab,
        gender;
        MISSING ARE .;
        WITHIN ARE gevocab gender;
        CLUSTER = school classID;

ANALYSIS: TYPE = THREELEVEL RANDOM;

MODEL:    %BETWEEN school%
             gender_sl;

          %BETWEEN classID%
           gender_sl;

          %WITHIN%
           geread on gevocab;
           gender_sl | geread on gender;
```

This syntax allows the gender coefficient to vary at both the school and classroom levels. The resulting output appears as follows:

```
Information Criteria

            Akaike (AIC)                    45209.159
            Bayesian (BIC)                  45267.450
            Sample-Size Adjusted BIC        45242.027
              (n* = (n + 2) / 24)
```

MODEL RESULTS

	Estimate	S.E.	Est./S.E.	Two-Tailed P-Value
Within Level				
GEREAD ON				
GEVOCAB	0.506	0.014	35.825	0.000
Residual Variances				
GEREAD	3.733	0.141	26.532	0.000
Between CLASSID Level				
Variances				
GEREAD	0.092	0.026	3.489	0.000
GENDER_SL	0.023	0.043	0.535	0.592

```
Between SCHOOL Level

  Means
       GEREAD            2.050      0.064     32.156      0.000
       GENDER_SL         0.018      0.038      0.475      0.635

  Variances
       GEREAD            0.073      0.017      4.373      0.000
       GENDER_SL         0.018      0.026      0.662      0.508
```

Interpreting these results, we first see that there is not a statistically significant relationship between the fixed gender effect and the reading achievement ($\gamma = 0.018$, $p = .635$). In other words, across classrooms and schools, the difference in mean reading achievement for males and females is not shown to be statistically significant, when accounting for vocabulary scores. The estimate for the gender random coefficient term is not significant at the school level ($\tau = 0.018$, $p = .508$) or the classroom level ($\tau = 0.023$, $p = .592$).

As noted previously, in model 4.5, the coefficients for gender were allowed to vary randomly across both classes and schools. However, there may be some situations in which a researcher is interested in allowing the coefficient for a fixed effect to vary for only one of the random effects, such as classroom, for example. We can very easily change the syntax for this model to allow gender to only vary at the classroom level. For example, the following code will allow gender to vary across classrooms but not schools:

```
ANALYSIS: TYPE = THREELEVEL RANDOM;

MODEL:     %BETWEEN school%
              geread

           %BETWEEN classID%
           gender_sl;

           %WITHIN%
           geread on gevocab;
           gender_sl | geread on gender;
```

Or we could have gender vary across schools but not classrooms.

```
ANALYSIS: TYPE = THREELEVEL RANDOM;

MODEL:     %BETWEEN school%
              gender_sl;

           %BETWEEN classID%
              geread;

           %WITHIN%
           geread on gevocab;
           gender_sl | geread on gender;
```

Interactions and Cross-Level Interactions in Three-Level Models

Now if we recall from Chapter 3, we can include both single-level and cross-level interactions in the model. Inclusion of single-level interactions is very simple and can be done easily for either TYPE = THREELEVEL or TYPE = THREELEVEL RANDOM by just using the DEFINE command to create a new variable for the interaction. For example, building on model 5.2, we could conceptualize a three-level random intercept model in which, in addition to vocabulary, class size, and school size variables, we are interested in whether there is an interaction between a child's vocabulary and their gender.

```
TITLE: 5.4 Three Level Random Intercepts Model with Level 1
Interaction
DATA: FILE IS Achieve_ch3.csv;
VARIABLE: NAMES ARE corp school class classID gender age
geread gevocab clenroll senroll;
          USEVARIABLES ARE school, classID, gender, geread,
          gevocab, clenroll, senroll, GV;
          WITHIN ARE gevocab gender GV;
          BETWEEN ARE (classID) clenroll (school) senroll;
          CLUSTER = school classID;

DEFINE:   Center gevocab gender (grandmean);
          GV = gevocab*gender;


ANALYSIS: TYPE = THREELEVEL;

MODEL:    %BETWEEN school%
                geread on senroll;

          %BETWEEN classID%
           geread on clenroll;

          %WITHIN%
           geread on gevocab gender GV;
```

As can be seen in the syntax, the DEFINE: command was used to create the new variable GV as the product (interaction) between gender and gevocab. When creating variables using the DEFINE: command, newly created variables should be added to the end of the USEVARIABLES line and should also appear on the proper level of analysis (here, WITHIN). Also, due to potential multicollinearity issues with the introduction of interaction terms, the variables of gevocab and gender was grand mean centered, which is accomplished in the DEFINE command as well.

```
Information Criteria

          Akaike (AIC)                    45204.937
          Bayesian (BIC)                  45270.514
          Sample-Size Adjusted BIC        45241.913
            (n* = (n + 2) / 24)
```

MODEL RESULTS

	Estimate	S.E.	Est./S.E.	Two-Tailed P-Value
Within Level				
GEREAD ON				
GEVOCAB	0.505	0.014	35.486	0.000
GENDER	0.018	0.038	0.486	0.627
GV	0.018	0.020	0.904	0.366
Residual Variances				
GEREAD	3.741	0.141	26.588	0.000
Between CLASSID Level				
GEREAD ON				
CLENROLL	0.023	0.009	2.534	0.011
Residual Variances				
GEREAD	0.096	0.026	3.644	0.000
Between SCHOOL Level				
GEREAD ON				
SENROLL	0.000	0.000	-0.551	0.581
Intercepts				
GEREAD	3.912	0.209	18.699	0.000
Residual Variances				
GEREAD	0.079	0.017	4.653	0.000

We will not go too deeply into the interpretation of this model; however, looking at the within level, we can see that GV (the interaction between gender and gevocab) is not significant ($\gamma = 0.018$, $p = .366$), indicating that gender does not change or moderate the relationship between vocabulary and reading. Gender itself is also not a significant predictor ($\gamma = 0.018$, $p = .627$), but like the previous models, gevocab is still a significant positive predictor of reading ($\gamma = 0.505$, $p < .001$).

The DEFINE command can be used very easily to create single-level inter-actions at any level of analysis for either random intercept (THREELEVEL) or random coefficients (THREELEVEL RANDOM) models. However, if, instead of single-level interactions, we wish to have variables across levels interact, we must always work from the THREELEVEL RANDOM command and the proce-dure is similar to that defined in Chapter 3 for two-level models. For example, we may have a hypothesis stating that the impact of vocabulary score on reading achievement varies depending upon the size of the school that a student attends. In order to test this hypothesis, we will need to include the interaction between vocabulary score and size of the school, as is done in model 5.5 as follows:

```
TITLE: 5.5 Three Level Model with Cross Level interaction
DATA: FILE IS Achieve_ch3.csv;
VARIABLE: NAMES ARE corp school class classID gender age
geread gevocab clenroll senroll;
          USEVARIABLES ARE school, classID, geread, gevocab,
          clenroll, senroll;
          WITHIN ARE gevocab;
          BETWEEN ARE (school) senroll (classID) clenroll;
          CLUSTER = school classID;

ANALYSIS: TYPE = THREELEVEL RANDOM;

MODEL:    %BETWEEN school%
            geread gevocab_sl on senroll;

          %BETWEEN classID%
           geread on clenroll;

          %WITHIN%
           gevocab_sl | geread on gevocab;
```

Recall from Chapter 3 that in order to model cross-level interactions, the lower level variable must first be defined as having a random slope (here this is occurring at the within-subjects level), and then this random slope is being explained by predictor(s) at a higher level (here this happens at the between-school level where the gevocab slope is being explained by school enrolment).

```
Information Criteria

          Akaike (AIC)                      45065.227
          Bayesian (BIC)                    45138.091
          Sample-Size Adjusted BIC          45106.313
             (n* = (n + 2) / 24)
```

MODEL RESULTS

	Estimate	S.E.	Est./S.E.	Two-Tailed P-Value
Within Level				
Residual Variances				
GEREAD	3.639	0.138	26.304	0.000
Between CLASSID Level				
GEREAD ON				
CLENROLL	-0.011	0.011	-1.056	0.291
Variances				
GEVOCAB_SL	0.007	0.001	5.126	0.000
Residual Variances				
GEREAD	0.001	0.047	0.031	0.975
Between SCHOOL Level				
GEVOCAB_SL ON				
SENROLL	0.000	0.000	-1.083	0.279
GEREAD ON				
SENROLL	0.000	0.000	1.084	0.278
Intercepts				
GEREAD	2.046	0.305	6.704	0.000
GEVOCAB_SL	0.562	0.051	11.056	0.000
Residual Variances				
GEREAD	0.054	0.028	1.920	0.055
GEVOCAB_SL	0.003	0.001	2.247	0.025

In terms of results, we will first be interested in whether or not the model, including the interaction, provides better fit to the data than model 5.1 with no interaction. And once again, we will make this decision based upon the AIC and BIC values. Given that these information indices are larger for model 5.1, we would conclude that including the interaction between vocabulary score and school size does yield a better fitting model. In terms of hypothesis testing results, we find that student vocabulary ($t = 11.056$, $p < .001$) is the only statistically significant predictor of reading ability.

Summary

This chapter is very much an extension of Chapter 4, extending the use of Mplus in fitting two-level models to include data structure at three levels. In practice, such complex multilevel data are relatively rare. However, as we saw in this chapter, faced with this type of data, we can use Mplus to model it appropriately. Indeed, the basic framework that we employed in the two-level case works in a very similar manner for the more complex data featured in this chapter. If you have read the first five chapters, you should now feel fairly comfortable analyzing most common multilevel models with a continuous outcome variable. We will next turn our attention to the application of multilevel models to longitudinal data. Of key importance as we change directions a bit is that the core ideas that we have already learned, including fitting of the null, random intercept, and random coefficients models, as well as inclusion of predictors at different levels of the data, do not change when we have longitudinal data. As we will see, application of multilevel models in this context is no different from that in Chapters 4 and 5. What is different is the way in which we define the levels of the data. Heretofore, level 1 has generally been associated with individual people. With longitudinal data, however, level 1 will refer to a single measurement in time, whereas level 2 will refer to the individual subject. By recasting longitudinal data in this manner, we make available to ourselves the flexibility and power of multilevel models.

6

Longitudinal Data Analysis Using Multilevel Models

To this point, we have focused on multilevel models in which a single measurement is made on each individual in the sample, and these individuals are in turn clustered. However, as mentioned in Chapter 2, multilevel modeling can be used in a number of different contexts, with varying data structures. In particular, this chapter will focus on using the multilevel modeling framework to analyze longitudinal data. Longitudinal data occurs when a series of measurements are made on each individual in the sample, usually over some period of time. Although there are alternatives to this temporal definition of longitudinal data (e.g., measurements made at multiple locations within a plot of land), we will focus on the most common type of longitudinal, which is time based. In this chapter, we will demonstrate the application to this special case of tools that we have already discovered. We will then conclude the chapter by describing the advantages of using multilevel models with longitudinal data.

The data to be used in the following examples come from the realm of educational testing. Examinees were given a language assessment at six equally spaced times. This file (`LanguagePP.csv`) includes the total language achievement test score measured over six different measurement occasions (the outcome variable), four language subtest scores (Writing process and features, Writing applications, Grammar, and Mechanics), and variables indicating student ID and school ID.

The Multilevel Longitudinal Framework

As with the two- and three-level multilevel models described in Chapters 3 and 5, longitudinal analysis in the multilevel framework involves regression-like equations at each level of the data. In the case of longitudinal models, the data structure takes the following form: repeated measurements (level 1) nested within the individual (level 2), and possibly individual nested within some higher level cluster (e.g., school) at level 3. A simple two-level

longitudinal model involving repeated measurements nested within individuals can be expressed as follows:
Level 1:

$$Y_{it} = \pi_{0i} + \pi_{1i}(T_{it}) + \pi_{2i}(X_{it}) + \varepsilon_{it} \tag{6.1}$$

Level 2:

$$\pi_{0i} = \beta_{00} + \beta_{01}(Z_i) + r_{0i}$$

$$\pi_{1i} = \beta_{10} + r_{1i}$$

$$\pi_{2i} = \beta_{20} + r_{2i}$$

where:
  Y_{it} is the outcome variable for individual i at time t
  π_{it} are the level 1 regression coefficients
  β_{it} are the level 2 regression coefficients
  ε_{it} is the level 1 error
  r_{it} are the level 2 random effects
  T_{it} is a dedicated time predictor variable
  X_{it} is a time-varying predictor variable
  Z_i is a time-invariant predictor

Thus, as can be seen in Equation 6.1, although there is new notation to define specific longitudinal elements, the basic framework for the multilevel model is essentially the same as we saw for the two-level model in Chapter 3. The primary difference is that now we have three different types of predictors: a time predictor, time-varying predictors, and time-invariant predictors. Given their unique role in longitudinal modeling, it is worth spending just a bit of time defining each of these predictor types.

Of the three types of predictors that are possible in longitudinal models, a dedicated time variable is the only one that is necessary in order to make the multilevel model longitudinal. This time predictor, which is literally an index of the time point at which a particular measurement was made, can be very flexible with time measured in fixed intervals or in waves. If time is measured in waves, they can be waves of varying length from person to person, or they can be measured on a continuum. It is important to note that when working with time as a variable, it is often worthwhile to rescale it so that the first measurement occasion is the zero point, thereby giving the intercept the interpretation of baseline, or initial status on the dependent variable.

The other two types of predictors (time varying and time-invariant) differ in terms of how they are measured. A predictor is time varying when it is measured at multiple points in time, just as is the outcome variable.

In the context of education, a time-varying predictor might be the number of hours in the previous 30 days a student has spent studying. This value could be recorded concurrently with the student taking the achievement test serving as the outcome variable. However, a predictor is time invariant when it is measured at only one point in time, and its value does not change across measurement occasions. An example of this type of predictor would be gender, which might be recorded at the baseline measurement occasion, and which is unlikely to change over the course of the data collection period. In the context of applying multilevel models to longitudinal data problems, time-varying predictors will appear at level 1 because they are associated with specific measurements, whereas time-invariant predictors will appear at level 2 or higher, because they are associated with the individual (or a higher data level) across all measurement conditions.

Person Period Data Structure

The first step in fitting multilevel longitudinal models with any statistical package is to make sure the data are in the proper longitudinal structure. Oftentimes, such data are entered in what is called person-level data structure. Person-level data structure includes one row for each individual in the dataset and one column for each variable or measurement on that individual. In the context of longitudinal data, this would mean that each measurement in time would have a separate column of its own. Although the person-level data structure works well in many instances, in order to apply multilevel modeling techniques to longitudinal analyses, the data must be reformatted into what is called person period data structure. In this format, rather than having one row for each individual, person period data has one row for each time that each subject is measured, so that data for an individual in the sample will consist of as many rows as there were measurements made.

The data to be used in the following examples includes the total language achievement test score measured over six different measurement occasions (the outcome variable), four language subtest scores (Writing process and features, Writing applications, Grammar, and Mechanics), and variables indicating student ID and school ID. Before running this data as a longitudinal multilevel model using Mplus, the data must be restructured from a person-level into a person period format. Restructuring person-level data into a person period format can be accomplished in a number of statistical packages. This restructure can be accomplished in R by using the stack command, as is demonstrated in the appendix.

Fitting Longitudinal Models Using Mplus

Once data have been restructured into a person period format, fitting longitudinal models in a multilevel framework can be done in exactly the same manner as we saw in Chapters 3 and 5. As we noted before, the primary difference between the scenario described here and that appearing in Chapters 3 and 5 is that the nesting structure reflects repeated measurements for each individual. For example, using the Language data we described in Chapter 3, we could use the following syntax for a longitudinal random intercept model predicting Language over time.

```
TITLE:6.1 Unconditional Growth Model
DATA: File IS LanguagePP.csv;
VARIABLE: NAMES ARE ID school Process Application Grammar Goal
Time Language;
          USEVARIABLES ARE ID Time Language;
          MISSING ARE .;
          WITHIN ARE Time;
          CLUSTER = ID;

ANALYSIS: TYPE = TWOLEVEL RANDOM

MODEL: %WITHIN%
          Language ON Time;
```

The following coefficient output would result.

```
Information Criteria
              Akaike (AIC)                    135076.629
              Bayesian (BIC)                  135107.869
              Sample-Size Adjusted BIC        135095.158
                (n* = (n + 2) / 24)
```

MODEL RESULTS

	Estimate	S.E.	Est./S.E.	Two-Tailed P-Value
Within Level				
LANGUAGE ON				
TIME	3.245	0.044	74.208	0.000
Residual Variances				
LANGUAGE	56.636	1.095	51.742	0.000
Between Level				
Means				
LANGUAGE	193.990	0.376	516.475	0.000
Variances				
LANGUAGE	232.042	5.510	42.114	0.000

Given that we have devoted substantial time in Chapters 3 and 5 interpreting multilevel model output, we will not spend a great deal of time here for that purpose. However, it is important to point out that these results indicate a statistically significant positive relationship between time and performance on the language assessment, such that scores increased over time.

Adding predictors to the model also remains the same as in earlier examples, regardless of whether they are time varying or time-invariant. For example, in order to add `Grammar`, which is time varying, as a predictor of total language scores over time we would use the following:

```
TITLE: 6.2 Change model with Time Varying Predictor;
DATA: File IS LanguagePP.csv;
VARIABLE: NAMES ARE ID school Process Application Grammar Goal
Time Language;
          USEVARIABLES ARE ID Grammar Time Language;
          MISSING ARE .;
          WITHIN ARE Grammar Time;
          CLUSTER = ID;

ANALYSIS: TYPE = TWOLEVEL;

MODEL: %WITHIN%
       Language ON Grammar Time;
```

With the resulting output:

```
Information Criteria
```

Akaike (AIC)	130015.197
Bayesian (BIC)	130054.248
Sample-Size Adjusted BIC	130038.358
(n* = (n + 2) / 24)	

```
MODEL RESULTS
```

	Estimate	S.E.	Est./S.E.	Two-Tailed P-Value
Within Level				
LANGUAGE ON				
GRAMMAR	0.631	0.006	106.319	0.000
TIME	3.245	0.044	74.211	0.000
Residual Variances				
LANGUAGE	56.640	1.095	51.739	0.000
Between Level				
Means				
LANGUAGE	70.460	1.174	60.009	0.000
Variances				
LANGUAGE	36.103	1.264	28.560	0.000

From these results, we see that again, Time is positively related to scores on the language assessment, indicating that they increased over time. In addition, Grammar is also statistically significantly related to language test scores, meaning that measurement occasions in with higher Grammar scores also had higher language scores. Finally, the lower Akaike's information criterion (AIC) and Bayesian information criterion (BIC) values for the model including Grammar than the one excluding it indicate that the former is a better fit.

If we wanted to allow the growth rate to vary randomly across individuals, we would use the following syntax:

```
TITLE:6.3 Random Slopes Model
DATA: File IS LanguagePP.csv;
VARIABLE: NAMES ARE ID school Process Application Grammar Goal
Time Language;
          USEVARIABLES ARE ID Time Language;
          MISSING ARE .;
          WITHIN ARE Time;
          CLUSTER = ID;

ANALYSIS: TYPE = TWOLEVEL RANDOM

MODEL: %WITHIN%
       Time_sl | Language ON Time;

       %BETWEEN%
       Time_sl;
```

With the corresponding output:

```
Information Criteria

            Akaike (AIC)                    134910.078
            Bayesian (BIC)                  134949.128
            Sample-Size Adjusted BIC        134933.238
              (n* = (n + 2) / 24)
```

MODEL RESULTS

	Estimate	S.E.	Est./S.E.	Two-Tailed P-Value
Within Level				
Residual Variances				
LANGUAGE	50.247	1.184	42.443	0.000
Between Level				
Means				
LANGUAGE	193.986	0.376	516.452	0.000
TIME_SL	3.244	0.044	74.184	0.000
Variances				
LANGUAGE	265.450	7.955	33.370	0.000
TIME_SL	1.317	0.157	8.391	0.000

In this model, the random effect for Time is assessing the extent to which growth over time differs from one person to the next. Results show that the random effect for Time (TIME_SL) is statistically significant. Thus, we can conclude that growth rates in language scores over the six time points do differ across individuals in the sample.

We could add a third level of data structure to this model by including information regarding schools, within which examinees are nested. To fit this model, we use the following syntax:

```
TITLE:6.4 Simple Three Level Random Coefficients Longitudinal
Model
   DATA: File IS LanguagePP.csv;
   VARIABLE: NAMES ARE ID school Process Application Grammar
Goal Time Language;
              USEVARIABLES ARE Language Time school ID;
              MISSING ARE .;
              CLUSTER = school ID;
              WITHIN = Time;

ANALYSIS: TYPE = THREELEVEL RANDOM;

MODEL: %BETWEEN school%
          Time_sl;

          %BETWEEN ID%
          Time_sl;

          %WITHIN%
          Time_sl | Language ON Time;
```

With the resulting output:

```
Information Criteria

              Akaike (AIC)               133704.497
              Bayesian (BIC)             133759.167
              Sample-Size Adjusted BIC   133736.922

              (n* = (n + 2) / 24)
```

MODEL RESULTS

	Estimate	S.E.	Est./S.E.	Two-Tailed P-Value
Within Level				
Residual Variances				
LANGUAGE	50.261	3.221	15.602	0.000

```
Between ID Level

  Variances
     LANGUAGE          197.989      13.213      14.985       0.000
     TIME_SL             0.608       0.228       2.669       0.008

Between SCHOOL Level

  Means
     LANGUAGE          194.672       2.187      89.005       0.000
     TIME_SL             3.098       0.242      12.789       0.000

  Variances
     LANGUAGE          146.264      30.101       4.859       0.000
     TIME_SL             1.715       0.510       3.364       0.001
```

Given that the AIC for Model 6.4 is lower than that for Model 6.3, where school is not included as a variable, we can conclude that inclusion of the school level of the data leads to better model fit.

Recall from Chapter 3 that when random effects are employed, it is possible (and often desirable) to estimate the relationship between the random intercept and the random slope(s). Especially when estimating longitudinal models where the time slope is allowed to vary, estimation and interpretation of these covariances is often desired. As mentioned in Chapter 3, Mplus by default sets these covariances to zero; however, it is possible, to override this default by specifying the correlation in the model syntax.

```
INPUT INSTRUCTIONS

TITLE:6.3 Random Slopes Model
  DATA: File IS LanguagePP.csv;
  VARIABLE: NAMES ARE ID school Process Application Grammar
Goal Time Language;
             USEVARIABLES ARE ID Time Language;
             MISSING ARE .;
             WITHIN ARE Time;
             CLUSTER = ID;

  ANALYSIS: TYPE = TWOLEVEL RANDOM

  MODEL: %WITHIN%
         Time_sl | Language ON Time;

         %BETWEEN%
         Language with Time_sl;
```

With the resulting output:

```
Information Criteria

          Akaike (AIC)                    133507.576
          Bayesian (BIC)                  133554.436
          Sample-Size Adjusted BIC        133535.369
            (n* = (n + 2) / 24)
```

```
MODEL RESULTS
```

	Estimate	S.E.	Est./S.E.	Two-Tailed P-Value
Within Level				
Residual Variances				
LANGUAGE	45.391	0.976	46.492	0.000
Between Level				
LANGUAGE WITH				
TIME_SL	-27.803	1.047	-26.545	0.000
Means				
LANGUAGE	193.983	0.376	516.418	0.000
TIME_SL	3.245	0.044	74.212	0.000
Variances				
LANGUAGE	389.129	8.850	43.967	0.000
TIME_SL	3.212	0.169	18.954	0.000

We can see that the correlation between the random intercept and the random slope (LANGUAGE WITH TIME_SL) is significant and negative, indicating that the higher the initial status (intercept), the less steep the slope for time (the random slope).

Benefits of Using Multilevel Modeling for Longitudinal Analysis

Modeling longitudinal data in a multilevel framework has a number of advantages over more traditional methods of longitudinal analysis (e.g., ANOVA designs). For example, using a multilevel approach allows for the simultaneous modeling of both intraindividual change (how an individual changes over time) and interindividual change (differences in this temporal change across individuals). A particularly serious problem that afflicts many longitudinal studies is high attrition within the sample. Quite often, it is difficult for researchers to keep track of members of the sample over time, especially over a lengthy period of time. When using traditional techniques for longitudinal data analysis such as repeated measures ANOVA, only complete data cases can be analyzed. Thus, when there is a great deal of missing data, either a sophisticated missing data replacement method (e.g., multiple imputation) must be employed, or the researcher must work with a greatly reduced sample size. In contrast, multilevel modeling is able to use the available data from incomplete observations, thereby not reducing the sample size as dramatically as other approaches for modeling longitudinal data, nor requiring special missing data methods. Repeated measures ANOVA is traditionally one of the most common methods for analysis of change. However, when used with longitudinal data, the assumptions upon which repeated measures rests may be too restrictive. In particular, the assumption of sphericity (assuming equal variances of outcome variable differences) may be unreasonable, given that variability can change considerably as time passes. However, analyzing longitudinal data from a multilevel modeling perspective does not require the assumption of sphericity. In addition, it also provides flexibility in model definition, thus allowing for information about the anticipated effects of time on error variability to be included in the model design. Finally, multilevel models can easily incorporate predictors from each of the data levels, thereby allowing for more complex data structures. In the context of longitudinal data, this means that it is possible to incorporate measurement occasion (level 1), individual (level 2), and cluster (level 3) characteristics. However, in the context of repeated measures ANOVA or multivariate analysis of variance (MANOVA), incorporating these various levels of the data would be much more difficult. Thus, use of multilevel modeling in this context not only has the benefits listed previously pertaining specifically to longitudinal analysis but also brings the added capability of simultaneous analysis of multiple levels of influence.

Summary

In this chapter, we saw that the multilevel modeling tools we studied together in Chapters 2 through 5 can be applied in the context of longitudinal data. The key to this analysis is the treatment of each measurement in time as a level 1 data point and the individuals on whom the measurements are made as level 2. Once this shift in thinking is made, the methodology remains very much the same as what we employed in the standard multilevel models in Chapters 3 and 4. By modeling longitudinal data in this way, we are able to incorporate a wide range of data structures, including individuals (level 2) nested within a higher level of data (level 3).

7

Brief Introduction to Generalized Linear Models

Heretofore, we have focused our attention primarily on models for data in which the outcome variables are continuous in nature. Indeed, we have been even more specific and dealt almost exclusively with models resting on the assumption that the model errors are normally distributed. However, in many applications, the outcome variable of interest is categorical, rather than continuous. For example, a researcher might be interested in predicting whether or not an incoming freshman is likely to graduate from college in 4 years, using high-school grade point average and admissions test scores as the independent variables. Here, the outcome is the dichotomous variable graduation in 4 years (yes or no). Likewise, consider research conducted by a linguist who has interviewed terminally ill patients and wants to compare the number of times those patients use the word death or dying during the interviews. The number of times that each word appears, when compared to the many thousands of words contained in the interviews is likely to be very small, if not zero for some people. In short, the dependent variable is the rate of target words used by the interviewees. Again, this rate will likely be very low, so that the model errors are almost assuredly not normally distributed. Yet another example of categorical outcome variables would occur when a researcher is interested in comparing scores by treatment condition on mathematics performance outcome that are measured on a Likert scale, such as 1, 2, or 3, where higher scores indicate better performance on the mathematics task. Thus, the multilevel models that we described in Chapters 2–5 would not be applicable to these research scenarios.

In each of the previous examples, the outcome variable of interest is not measured on a continuous scale, and will very likely not produce normally distributed model errors. As we have seen, the linear multilevel models discussed previously operate under the assumption of normality of errors. As such, they will not be adequate for situations in which these or other types of variables that cannot be appropriately modeled with a linear model are to be used. However, alternative models for such variables are available. Taken together, these alternatives for categorical outcome variables are often referred to as generalized linear models (GLiMs). Prior to discussing the multilevel versions of these models in Chapter 8, it will behoove us to first explore some common GLiMs and their applications in the single-level context. In the next chapter, we will then expand upon our discussion here to include the multilevel variants of these models, and how to fit them in Mplus. In the following sections of this chapter, we will focus on three broad types of GLiMs, including those for categorical outcomes (dichotomous, ordinal, and nominal), counts or rates of events that occur very infrequently, and counts or rates of events that occur somewhat more frequently. After their basic theoretical presentation, we will then describe how these single-level GLiMs can be fit using functions in Mplus.

Logistic Regression Model for a Dichotomous Outcome Variable

As an example of a GLiM, we begin the discussion of models for dichotomous outcome data. Let us consider an example involving a sample of 20 men, 10 of whom have been diagnosed with coronary artery disease, and 10 who have not. Each of the 20 individuals was asked to walk on a treadmill until he became too fatigued to continue. The outcome variable in this study was the diagnosis and the independent variable was the time walked until fatigue, that is, the point at which the subject requested to stop. The goal of the study was to find a model predicting coronary artery status as a function of time walked until fatigue. If an accurate predictive equation could be developed, it might be a helpful tool for physicians to use in diagnosing heart problems in patients. In the context of Chapter 1, we might consider applying a linear regression model to these data, as we found that this approach was useful for estimating predictive equations. However, recall that there were a number of assumptions on which appropriate inference in the context of linear regression depends, including that the residuals would be normally distributed. Given that the outcome variable in the current problem is a dichotomy (coronary disease or not), the residuals will almost certainly not follow a normal distribution. Therefore, we need to identify an alternative approach for dealing with dichotomous outcome data such as these.

Perhaps the most commonly used model for linking a dichotomous outcome variable with one or more independent variables (either continuous or categorical) is logistic regression. The logistic regression model takes the following form:

$$\ln\left[\frac{p(y=1)}{1-p(y=1)}\right] = \beta_0 + \beta_1 x \qquad (7.1)$$

Here, y is the outcome variable of interest, taking the values 1 or 0, where 1 is typically the outcome of interest. (Note that these dichotomous outcomes could also be assigned other values, though 1 and 0 are probably the most commonly used in practice). This outcome is linked to an independent variable, x, by the slope (β_1) and intercept (β_0). Indeed, the right side of this equation should look very familiar, as it is identical to the standard linear regression model. However, the left side is quite different from what we see in linear regression, due to the presence of the logistic link function, also known as the logit. Within the parentheses lie the odds that the outcome variable will take the value of 1. For our coronary artery example, 1 is the value for having coronary artery disease and 0 is the value for not having it. In order to render the relationship between this outcome and the independent variable (time walking on treadmill until fatigue) linear, we need to take the natural log of these odds. Thus, the logit link for this problem is the natural log of the odds of an individual having coronary artery disease. Interpretation of the slope and intercept in the logistic regression model are the same as interpretation in the linear regression context. A positive value of β_1 would indicate that the larger the value of x, the greater the log odds of the target outcome occurring. The parameter β_0 is the log odds of the target event occurring when the value of x is 0. Logistic regression models can be fit easily in Mplus. In the following section, we will see an example program, and how to interpret the results we obtain from it.

The data were read from a data set called `coronary_chapter_7.dat`, which appears as follows:

```
1014    0
 684    0
 810    0
 990    0
 840    0
 978    0
1002    0
1110    0
 864    1
 636    1
 638    1
 708    1
 786    1
 600    1
 720    1
 750    1
 594    1
 750    1
1112    0
1359    0
```

The first column contains the number of seconds that patients could walk prior to becoming fatigued, and the second column indicates whether the individual has coronary artery disease (1) or not (0). The Mplus program `model_7.1`, which appears next, can be used to conduct the logistic regression analysis relating these two variables to one another.

```
TITLE:      Model 7.1 Coronary artery logistic regression model
DATA:       FILE IS coronary_chapter_7.dat;
VARIABLE:   NAMES ARE id time disease;
            CATEGORICAL IS disease;
         usevariables are time disease;
ANALYSIS:   ESTIMATOR = ML;
MODEL:      disease ON time;
```

As with all Mplus programs, it is helpful to include a title explaining the purpose of this analysis. The data file is then called in the DATA: command line. The dataset contains three variables, id, time, and disease, of which we will treat disease as categorical. This last point is crucial here because we must define disease as CATEGORICAL, in order for Mplus to treat it correctly in the subsequent MODEL: command, that is, with logistic regression. In addition, we need to be sure to include only the two variables of interest, time and disease, in the analysis, and thus exclude the subject id number. Finally, we fit the model using maximum likelihood, and define the regression model just as we do in a standard regression analysis with dependent variable ON independent variable.

The output appears in its own window, just as was the case for all of the other analyses that we have worked on thus far. When interpreting logistic regression results, it is important to know which of the two possible outcomes is being modeled by the software in the numerator of the logit. In other words, we need to know which category was defined as the target by the software so that we can properly interpret the model parameter estimates. By default, Mplus will treat the higher value as the target. In this case, 0 = healthy and 1 = disease. Therefore, the numerator of the logit will be 1, or disease. This is a very important consideration, as the results would be completely misinterpreted if the user is unclear on the numerator of the logit. The relevant results of the analysis appear as follows:

```
UNIVARIATE PROPORTIONS AND COUNTS FOR CATEGORICAL VARIABLES

    DISEASE
        Category 1      0.500            10.000
        Category 2      0.500            10.000


THE MODEL ESTIMATION TERMINATED NORMALLY


MODEL FIT INFORMATION

Number of Free Parameters                          2
```

```
Loglikelihood

        H0 Value                              -6.483

Information Criteria

        Akaike (AIC)                          16.966
        Bayesian (BIC)                        18.957
        Sample-Size Adjusted BIC              12.792
          (n* = (n + 2) / 24)



MODEL RESULTS

                                                       Two-Tailed
                      Estimate      S.E.   Est./S.E.    P-Value

DISEASE      ON
    TIME               -0.017      0.007     -2.247      0.025

Thresholds
    DISEASE$1         -13.489      5.877     -2.295      0.022


LOGISTIC REGRESSION ODDS RATIO RESULTS

DISEASE      ON
    TIME                0.984
```

Our initial interest is in determining whether there is a significant relationship between the independent variable (time) and the dependent (coronary disease status). Thus, we will first take a look at the row entitled TIME, as it contains this information. The column labeled estimate includes the slope and intercept values. The estimate of β_1 is −0.017, indicating that the more time an individual could walk on the treadmill before becoming fatigued, the lower the log odds that he or she had coronary artery disease, that is, the less likely he or she was to have heart disease. Through the simple transformation of the slope e^{β_1}, we can obtain the odds of having coronary artery disease as a function of time. This value appears at the very end of the output and is entitled LOGISTIC REGRESSION ODDS RATIO RESULTS. For this example, $e^{-0.017}$ is 0.984, indicating that for every additional second, an individual is able to walk on the treadmill before becoming fatigued, and their estimated odds of having heart disease are multiplied by 0.984. Thus, for an additional minute of walking, the odds decrease by $\exp(-.016534*60) = .378$.

Adjacent to the coefficient column is the standard error, which measures the sampling variation in the parameter estimate. The estimate divided by the standard error yields the test statistic, which appears under the Est./S.E.

column. This is the test statistic for the null hypothesis that the coefficient is equal to 0. Next is the p-value for the test statistic. Using standard practice, we would conclude that a p-value less than 0.05 indicates statistical significance. For this example, the p-value of 0.025 for time means that there is a statistically significant relationship between time on the treadmill to fatigue and the odds of an individual having coronary artery disease. The negative sign for the estimate further tells us that more time spent on the treadmill was associated with a lower likelihood of having heart disease. Finally in terms of the model terms, the row labeled threshold contains the model intercept, which in this case is −13.489. We would interpret this value as the log odds of an individual having heart disease if their walking time was 0. Clearly, this is not a very interesting piece of information, given that we would expect all but the most seriously ill individuals to be able to walk on the treadmill for at least one second.

In addition to the model parameter estimates, Mplus also provides the user with relative model fit indices, including the AIC, BIC, and aBIC statistics. We have defined these in Chapter 2, and thus will not do so again here. As we have seen in previous chapters, these values are useful for comparing the fit of different, and not necessarily nested, models, with smaller values indicating better relative fit to the data. Therefore, if we wanted to assess whether including additional independent variables or interactions improved the model fit, we could compare AIC, BIC, and aBIC values to ascertain which model was optimal. For the current example, there are no other independent variables of interest. However, it is possible to obtain the AIC for the intercept only model using the following program. The purpose behind doing so would be to determine whether including the time walking on the treadmill actually improved the model fit, after the penalty for model complexity was applied. The relevant output appears below the program.

```
TITLE:      Model 7.1 Coronary artery logistic regression model
DATA:       FILE IS coronary_chapter_7.dat;
VARIABLE:   NAMES ARE id time disease;
            CATEGORICAL IS disease;
       usevariables are disease;
ANALYSIS:   ESTIMATOR = ML;
MODEL:      disease ON ;


THE MODEL ESTIMATION TERMINATED NORMALLY



MODEL FIT INFORMATION

Number of Free Parameters                          1
```

```
Loglikelihood

        H0 Value                           -13.863

Information Criteria

        Akaike (AIC)                        29.726
        Bayesian (BIC)                      30.722
        Sample-Size Adjusted BIC            27.639
          (n* = (n + 2) / 24)
```

Chi-Square Test of Model Fit for the Binary and Ordered
Categorical
(Ordinal) Outcomes

```
        Pearson Chi-Square

        Value                               0.000
        Degrees of Freedom                      0
        P-Value                            1.0000

        Likelihood Ratio Chi-Square

        Value                               0.000
        Degrees of Freedom                      0
        P-Value                            1.0000
```

MODEL RESULTS

	Estimate	S.E.	Est./S.E.	Two-Tailed P-Value
Thresholds				
DISEASE$1	0.000	0.447	0.000	1.000

RESULTS IN PROBABILITY SCALE

DISEASE				
Category 1	0.500	0.112	4.472	0.000
Category 2	0.500	0.112	4.472	0.000

The AIC for this intercept only model was 29.726, which is larger than the 16.966 for the model including time. Based on AIC, along with the hypothesis test results discussed earlier, we would therefore conclude that the full model including Time provided better fit to the outcome of coronary artery disease.

Logistic Regression Model for an Ordinal Outcome Variable

In the prior example, we considered an outcome variable that could take two possible values, 0 (healthy heart), 1 (diseased heart). However, in many instances, a categorical outcome variable will have more than two potential outcomes. In this section, we demonstrate the case where the dependent variable is ordinal in nature so that the categories can be interpreted as going from less to more, or smaller to larger (or vice versa). In the next section, we will work with models that allow the categories to be unordered in nature.

As a way to motivate our discussion of ordinal logistic regression models, let us consider the following example. A dietician has developed a behavior management system for individuals suffering from obesity that is designed to encourage a healthier lifestyle. One such healthy behavior is the preparation of their own food at home using fresh ingredients rather than dining out or eating prepackaged foods. Study participants consisted of 100 individuals who were under a physician's care for a health issue directly related to obesity. Members of the sample were randomly assigned to either a control condition in which they received no special instruction in how to plan and prepare healthy meals from scratch, or a treatment condition in which they did receive such instruction. The outcome of interest was a rating provided 2 months after the study began, in which each subject indicated the extent to which he or she prepared his or her own meals. The response scale ranged from 0 (prepared all of my own meals from scratch) to 4 (never prepared any of my own meals from scratch), so that lower values were indicative of a stronger predilection to prepare meals at home from scratch. The dietician is interested in whether there are differences in this response between the control and treatment groups.

One commonly used model for such ordinal data is the cumulative logits model, which is expressed as follows:

$$
\text{logit}[P(Y \leq j)] = \ln \left[\frac{P(Y \leq j)}{1 - P(Y \leq j)} \right] \tag{7.2}
$$

In this model, there are $J-1$ logits where J is the number of categories in the dependent variable and Y is the actual outcome value. Essentially, this model compares the likelihood of the outcome variable taking a value of j or lower versus outcomes larger than j. For the current example, there would be four separate logits:

$$\ln\left[\frac{p(Y=0)}{p(Y=1)+p(Y=2)+p(Y=3)+p(Y=4)}\right]=\beta_{01}+\beta_1 x$$

$$\ln\left[\frac{p(Y=0)+p(Y=1)}{p(Y=2)+p(Y=3)+p(Y=4)}\right]=\beta_{02}+\beta_1 x$$

$$\ln\left[\frac{p(Y=0)+p(Y=1)+p(Y=2)}{p(Y=3)+p(Y=4)}\right]=\beta_{03}+\beta_1 x \tag{7.3}$$

$$\ln\left[\frac{p(Y=0)+p(Y=1)+p(Y=2)+p(Y=3)}{p(Y=4)}\right]=\beta_{04}+\beta_1 x$$

In the cumulative logits model, there is a single slope relating the independent variable to the ordinal response, and each logit has a unique intercept. In order to apply single slope across all logits, we must make the proportional odds assumption, which states that this slope is identical across logits. In order to fit the cumulative logits model to our data in Mplus, we use a very similar program to the one that we used for binary logistic regression. Mplus will count the number of categories for any variable that is specified to be categorical. In turn, when that variable is treated as the response in a regression model, Mplus will fit the model in 7.2 when more than two categories are present. The data for this example are contained in the file cooking_chapter_7.dat, and the program appears as follows:

```
TITLE:        Model 7.2 Cooking logistic regression model
DATA:         FILE IS cooking_chapter_7.dat;
VARIABLE:     NAMES ARE trt sex cook;
              CATEGORICAL IS cook;
       usevariables are trt cook;
ANALYSIS:     ESTIMATOR = ML;
MODEL:        cook ON trt;
```

The dependent variable, cook, must be specified as categorical, and the independent variable(s) cannot be categorical in nature. This latter fact means that categorical independent variables must be dummy coded, as are trt and sex in this case. The variable treatment is coded as 0 (control) and 1 (treatment). This program produces the following output:

```
THE MODEL ESTIMATION TERMINATED NORMALLY


MODEL FIT INFORMATION

Number of Free Parameters                    5
```

```
Loglikelihood

        H0 Value                                -146.567

Information Criteria

        Akaike (AIC)                            303.135
        Bayesian (BIC)                          316.161
        Sample-Size Adjusted BIC                300.369
          (n* = (n + 2) / 24)
```

MODEL RESULTS

		Estimate	S.E.	Est./S.E.	Two-Tailed P-Value
COOK	ON				
TRT		-0.796	0.368	-2.166	0.030
Thresholds					
COOK$1		-2.926	0.438	-6.678	0.000
COOK$2		-1.721	0.328	-5.254	0.000
COOK$3		-0.243	0.275	-0.882	0.378
COOK$4		1.373	0.323	4.253	0.000

LOGISTIC REGRESSION ODDS RATIO RESULTS

```
COOK        ON
    TRT                     0.451
```

BRANT WALD TEST FOR PROPORTIONAL ODDS

	Chi-Square	Degrees of Freedom	P-Value
COOK			
Overall test	0.000	3	1.000
TRT	0.000	3	1.000

The coefficient estimate for the treatment is −0.796, indicating that a higher value on the treatment variable (i.e., treatment = 1) was associated with a greater likelihood of providing a lower response on the cooking item.

Remember that lower responses to the cooking item reflected a greater propensity to eat scratch-made food at home. Thus, in this example, those in the treatment conditions had a greater likelihood of eating scratch-made food at home. Adjacent to the coefficient value is the standard error for the slope, which is divided into the coefficient in order to obtain the test statistic residing in the next column. The *p*-value associated with this test statistic is 0.03, which is less than the $\alpha = 0.05$ criterion. Given this result, we would conclude that there is indeed a statistically significant negative relationship between treatment condition and self-reported cooking behavior. Furthermore, by exponentiating the slope, we can also calculate the relative odds of a higher level response to the cooking item between the two groups. Much as we did in the dichotomous logistic regression case, we use the equation e^{β_1} to convert the slope to an odds ratio. In this case, the value is 0.451, indicating that the odds of a treatment group member selecting a higher level response (less self-cooking behavior) is only 0.451 as large as that of the control group. Note that this odds ratio applies to any pair of adjacent categories, such as 0 versus 1, 1 versus 2, 2 versus 3, or 3 versus 4. This value appears under the output line LOGISTIC REGRESSION ODDS RATIO RESULTS.

Mplus also provides the individual intercepts, as well as the AIC for the model. The intercepts are, as with dichotomous logistic regression, the log odds of the target response when the independent variable is 0. In this example, a treatment of 0 corresponds to the control group. Thus, the intercept represents the log odds of the target response for the control condition. As we saw earlier, it is possible to convert this to the odds scale through exponentiating the estimate. The first intercept provides the log odds of a response of 0 versus all other values for the control group, that is, plans and prepares all meals for oneself versus all other options. The intercept for this logit is −2.9259, which yields an $e^{-2.9259}$ of 0.054. We can interpret this to mean that the odds of a member of the control group planning and preparing their own meals versus something less are 0.054. In other words, it is highly unlikely a member of the control group will do this.

We can use the AIC value to compare the fit of this model with that of another model, just as we did for the binary logistic regression model. As an example, let us fit the null cumulative logits model including no independent variables.

```
TITLE:      Model 7.2b Cooking logistic regression model
DATA:       FILE IS cooking_chapter_7.dat;
VARIABLE:   NAMES ARE trt sex cook;
            CATEGORICAL IS cook;
        usevariables are cook;
ANALYSIS:   ESTIMATOR = ML;
MODEL:      cook ON ;
```

As with the binary logistic regression model, no independent variable is listed to the right of ON in the MODEL: line. The resulting output appears as follows:

```
THE MODEL ESTIMATION TERMINATED NORMALLY


MODEL FIT INFORMATION

Number of Free Parameters                          4

Loglikelihood

        H0 Value                           -148.957

Information Criteria

            Akaike (AIC)                    305.914
            Bayesian (BIC)                  316.334
            Sample-Size Adjusted BIC        303.701
              (n* = (n + 2) / 24)

Chi-Square Test of Model Fit for the Binary and Ordered
Categorical
(Ordinal) Outcomes

            Pearson Chi-Square

            Value                             0.000
            Degrees of Freedom                    0
            P-Value                          1.0000

            Likelihood Ratio Chi-Square

            Value                             0.000
            Degrees of Freedom                    0
            P-Value                          1.0000



MODEL RESULTS

                                                        Two-Tailed
                     Estimate      S.E.   Est./S.E.      P-Value

    Thresholds
        COOK$1        -2.442      0.369    -6.626         0.000
        COOK$2        -1.266      0.241    -5.243         0.000
```

COOK$3	0.160	0.201	0.799	0.424
COOK$4	1.735	0.280	6.194	0.000

RESULTS IN PROBABILITY SCALE

COOK

Category 1	0.080	0.027	2.949	0.003
Category 2	0.140	0.035	4.035	0.000
Category 3	0.320	0.047	6.860	0.000
Category 4	0.310	0.046	6.703	0.000
Category 5	0.150	0.036	4.201	0.000

The AIC for the null model is 305.914, which is larger than the 303.135 that we obtained for the model including treatment. Therefore, we can conclude that including treatment in the model improves the model fit.

Multinomial Logistic Regression

A third type of categorical outcome variable is one in which there are more than two categories but the categories are not ordered. An example can be seen in a survey of likely voters who were asked to classify themselves as liberal, moderate, or conservative. A political scientist might be interested in predicting an individual's political viewpoint as a function of his/her age. The most common statistical approach for doing so is the generalized logits or multinomial logistic regression model. This approach, which Agresti (2002) refers to as the baseline category logit model, assigns one of the dependent variable categories to be the baseline against which all other categories are compared. More formally, the multinomial logistic regression model can be expressed as follows:

$$\ln\left[\frac{p(Y=i)}{p(Y=j)}\right] = \beta_{i0} + \beta_{i1}x \qquad (7.4)$$

In this model, category j will always serve as the reference group against which the other categories, i, are compared. There will be a different logit for each nonreference category, and each of the logits will have a unique intercept (β_{i1}) and slope (β_{i1}). Thus, unlike the cumulative logits model in which a single slope represented the relationship between the independent variable and the outcome, in the multinomial logits model, we have multiple slopes for each independent variable, one for each logit. Therefore, we do not need to make the proportional odds assumption, which makes this model a useful alternative to the cumulative logits model when that assumption is not tenable. The disadvantage of using the multinomial logits model with an ordinal outcome variable is that the ordinal nature of the data is ignored. Any of the categories can serve as the reference, with the decision being based on the research question of most interest (i.e., against which group would comparisons be most interesting), or on pragmatic concerns such as which group is the largest, should the research questions not serve as the primary deciding factor. Finally, it is possible to compare the results for two nonreference categories using the following equation:

$$\ln\left[\frac{p(Y=i)}{p(Y=m)}\right] = \ln\left[\frac{p(Y=i)}{p(Y=j)}\right] - \ln\left[\frac{p(Y=m)}{p(Y=j)}\right] \tag{7.5}$$

For the present example, we will set the conservative group to be the reference, and fit a model in which age is the independent variable and political viewpoint is the dependent. The data file is `politics_chapter_7.dat` and contains the variables `age`, `view1`, and `view2`. There are two versions of the political viewpoint variable, with `view1` expressing it in numbers and `view2` in letters. Mplus only accepts numeric variables, so that in this case we will use `view1`. The categories are coded as 0 (liberal), 1 (moderate), or 2 (conservative) for each individual in the sample. Age was expressed in years. The Mplus program to fit the multinomial logistic regression model is as follows:

```
TITLE:      Model 7.3 Political viewpoint logistic regression
model
DATA:       FILE IS politics_chapter_7.dat;
VARIABLE:   NAMES ARE age view1 view2;
            nominal IS view1;
        usevariables are view1 age;
ANALYSIS:   ESTIMATOR = ML;
MODEL:      view1 ON age;
```

The results appear as follows:

```
UNIVARIATE PROPORTIONS AND COUNTS FOR CATEGORICAL VARIABLES

    VIEW1
        Category 1      0.257           386.000
        Category 2      0.389           584.000
        Category 3      0.353           530.000



THE MODEL ESTIMATION TERMINATED NORMALLY



MODEL FIT INFORMATION

Number of Free Parameters                                4
```

```
Loglikelihood

          H0 Value                              -1617.105

Information Criteria

          Akaike (AIC)                          3242.210
          Bayesian (BIC)                        3263.463
          Sample-Size Adjusted BIC              3250.756
            (n* = (n + 2) / 24)
```

MODEL RESULTS

		Estimate	S.E.	Est./S.E.	Two-Tailed P-Value
VIEW1#1	ON				
AGE		-0.017	0.004	-4.180	0.000
VIEW1#2	ON				
AGE		-0.005	0.003	-1.439	0.150
Intercepts					
VIEW1#1		0.440	0.191	2.298	0.022
VIEW1#2		0.330	0.172	1.911	0.056

LOGISTIC REGRESSION ODDS RATIO RESULTS

VIEW1#1	ON	
AGE		0.984
VIEW1#2	ON	
AGE		0.995

Based on these results, we see that the slope relating age to the logit comparing self-identification as liberal (VIEW#1) to conservative is −0.017, indicating that older individuals had a lower likelihood of being liberal versus conservative. In order to determine whether this relationship is statistically significant, we refer to the *p*-value, which is less than 0.05, leading us to conclude that the coefficient is statistically significant. In other words, we can conclude that in the population, older individuals are less likely to self-identify as liberal than as conservative. The hypothesis test result for moderate (VIEW#2) versus conservative was not statistically significant ($p = 0.150$). In other words, age is not related to the political viewpoint of an individual when it comes to comparing moderate versus conservative. Finally, we can calculate estimates for comparing the relationship of age with liberal versus moderate by applying Equation 7.5.

$$\ln\left[\frac{p(Y=0)}{p(Y=1)}\right] = \ln\left[\frac{p(Y=0)}{p(Y=2)}\right] - \ln\left[\frac{p(Y=1)}{p(Y=2)}\right]$$

$$= [0.440 - 0.017(\text{age})] - [0.330 - 0.005(\text{age})]$$

$$= -.110 - 0.0117(\text{age})$$

Taken together, we would conclude that older individuals are less likely to be liberal than conservative, and less likely to be liberal than moderate.

Models for Count Data

Poisson Regression

To this point, we have been focused on outcome variables of a categorical nature, such as whether an individual cooks for her/himself or the presence/absence of coronary artery disease. Another type of data that does not fit nicely into the standard models assuming normally distributed errors involves counts or rates of some outcome, particularly of rare events. Such variables often follow the Poisson distribution, a major property of which is that the mean is equal to the variance. It is clear that if the outcome variable is a count, its lower bound must be 0, that is, one cannot have negative counts. This presents a problem to researchers applying the standard linear regression model, as it may produce predicted values of the outcome that are less than 0, and thus are nonsensical. In order to deal with this potential difficulty, Poisson regression was developed. This approach to dealing with count data rests upon the application of the log to the outcome variable, thereby overcoming the problem of negative predicted counts, since the log of the outcome can take any real number value. Thus, when dealing with the Poisson distribution in the form of counts, we will use the log as the link function in fitting the Poisson regression model as follows:

$$\ln(Y) = \beta_0 + \beta_1 x \tag{7.6}$$

In all other respects, the Poisson model is similar to other regression models in that the relationship between the independent and dependent variables is expressed through the slope, β_1. And again, the assumption underlying the Poisson model is that the mean is equal to the variance.

This assumption is typically expressed by stating that the overdispersion parameter, $\phi = 1$. The ϕ parameter appears in the Poisson distribution density and thus is a key component in the fitting function used to determine the optimal model parameter estimates in maximum likelihood. A thorough review of this fitting function is beyond the scope of this book. Interested readers are referred to Agresti (2002) for a complete presentation of this issue.

Estimating the Poisson regression model using Mplus is quite straightforward, with the program being very similar to those that we have used for linear and logistic regression models. Consider an example in which a demographer is interested in determining whether there exists a relationship between the socioeconomic status (sei) of a family and the number of children under the age of 6 months (babies) who are living in the home. The data are in the file poisson_chapter_7.dat, and are read in and analyzed using the following Mplus program:

```
TITLE:       Model 7.4  Poisson regression model
DATA: FILE IS poisson_chapter_7.dat;
VARIABLE:    NAMES ARE babies sei;
             count IS babies;
         usevariables are babies sei;
MODEL:       babies ON sei;
```

The basic structure of the program is quite similar to that of the logistic regression models, with the primary difference being the inclusion of the count IS line, which specifies that the variable babies represents a count. The resulting output appears as follows:

```
COUNT PROPORTION OF ZERO, MINIMUM AND MAXIMUM VALUES

     BABIES       0.826          0          9




THE MODEL ESTIMATION TERMINATED NORMALLY




MODEL FIT INFORMATION

Number of Free Parameters                        2

Loglikelihood

          H0 Value                        -979.102
          H0 Scaling Correction Factor      2.1026
             for MLR
```

```
Information Criteria

          Akaike (AIC)                      1962.203
          Bayesian (BIC)                    1972.830
          Sample-Size Adjusted BIC          1966.476
             (n* = (n + 2) / 24)
```

```
MODEL RESULTS
```

		Estimate	S.E.	Est./S.E.	Two-Tailed P-Value
BABIES	ON				
SEI		-0.002	0.004	-0.362	0.718
Intercepts					
BABIES		-1.344	0.182	-7.384	0.000

These results show that sei did not have a statistically significant relationship with the number of children under 6 months old living in the home ($p = 0.718$).

The AIC, BIC, and aBIC will be useful as we compare the relative fit of the Poisson regression model with that of other models for count data.

Models for Overdispersed Count Data

Recall that a primary assumption underlying the Poisson regression model is that the mean and variance are equal. When this assumption does not hold, such as when the variance is larger than the mean, estimation of model standard errors is compromised so that they tend to be smaller than is actually true in the population (Agresti, 2002). For this reason, it is important that researchers dealing with count data investigate whether this key assumption is likely to hold in the population. An alternative to the Poisson model that can be used when the data are overdispersed is a regression model based on the negative binomial distribution. The mean of the negative binomial distribution is identical to that of the Poisson, whereas the variance is as follows:

$$\text{var}(Y) = \mu + \frac{\mu^2}{\theta} \tag{7.7}$$

From Equation 7.7, it is clear that as θ increases in size, the variance approaches the mean and the distribution becomes more like the Poisson. It is possible for the researcher to provide a value for θ if it is known that the data come from a particular distribution with a known θ. For example, when θ = 1, the data are modeled from the Gamma distribution. However, for most applications, the distribution is not known, in which case θ will be estimated from the data.

The negative binomial distribution can be fit to the data in Mplus using a slight variation to that in Model 7.4.

```
TITLE:      Model 7.5 Negative Binomial regression model
DATA:       FILE IS poisson_chapter_7.dat;
VARIABLE:   NAMES ARE babies sei;
            count IS babies (nb);
        usevariables are babies sei;
MODEL:      babies ON sei;
```

The key difference here between models 7.4 and 7.5 is that in this latter model, we indicate that the count follows the negative binomial distribution by including (nb) in the count line. The output appears as follows:

```
THE MODEL ESTIMATION TERMINATED NORMALLY



MODEL FIT INFORMATION

Number of Free Parameters                        3

Loglikelihood

        H0 Value                         -911.731
        H0 Scaling Correction Factor       1.4581
            for MLR

Information Criteria

        Akaike (AIC)                     1829.462
        Bayesian (BIC)                   1845.401
        Sample-Size Adjusted BIC         1835.871
            (n* = (n + 2) / 24)



MODEL RESULTS

                                                    Two-Tailed
                    Estimate      S.E.   Est./S.E.    P-Value

BABIES     ON
    SEI              -0.002      0.004      -0.360      0.719
```

Intercepts				
BABIES	-1.343	0.186	-7.212	0.000
Dispersion				
BABIES	2.474	0.511	4.839	0.000

The parameter estimates for the negative binomial regression are identical to those for the Poisson. This fact simply reflects the common mean that the distributions all share. However, the standard errors for the estimates can differ between the models, although in this case they are very similar. Indeed, the resulting hypothesis test results provide the same answer for both models, that there is not a statistically significant relationship between the sei and the number of babies living in the home. In terms of determining which model is optimal, we can compare the AIC, BIC, and aBIC from the negative binomial to those of the Poisson, with smaller values indicating better fit. The results here demonstrate that the negative binomial provides somewhat better fit to the data than does the Poisson. In short, it would appear that the data are somewhat overdispersed, as the model designed to account for this (negative binomial) provides better fit than the Poisson, which assumes no overdispersion. From a more practical perspective, the results of the two models are very similar, and a researcher using $\alpha = 0.05$ would reach the same conclusion regarding the lack of relationship between sei and the number of babies living in the home, regardless of which model he or she selected.

Introduction to Fitting Survival Models in Mplus

We will now discuss models designed for time until event data. Consider the situation in which a medical researcher is interested in what variables might be associated with the time until a particular outcome occurs, such as death. The researcher collects data on a sample of individuals beginning with baseline measurements, including whether he or she participated in a treatment designed to improve health, thereby hopefully delaying death, as well as their age at the beginning of the study. The subjects are then followed for 12 months, during which time their survival is recorded as the number of days until the event of interest (death). For those who died during the study period, the value of time is recorded as the number of days until they died. Those individuals who survive beyond the 12-month period are considered to be right censored, as the event of interest did not occur during the study period. Their value for time is coded as 365 because as of the last day of data collection, they were still alive. Other individuals may have moved away or simply stopped participating in the study. These individuals are also considered to be right censored, and their value of time is equal to their last of data collection. So, for example, if an individual moved away after 178 days in the study, their value for time would be recorded as 178. In addition to time, it is also important for the researcher to code whether each individual was censored or not. Thus, an additional variable is coded as 1 for those who died during the period of data collection and 0 for those who did not, and were therefore censored. Once the data are collected, the researcher would like to assess whether being in the treatment condition was associated with longer survival times, accounting for those who were censored, and whether there was a relationship between age at the beginning of the study and survival time.

There exist a wide variety of models for fitting time until event data, and addressing the types of research questions that are described earlier. Given that the purpose of this discussion is to introduce the reader unfamiliar with survival analysis data to the basic framework of these analyses, we will focus our attention on the most widely used of these methods—the Cox proportional hazards model. While it is certainly true that this model is very widely used and has proven to be quite flexible for many applications, it is also true that there are a number of other approaches that have proven useful in specific situations. One of the primary advantages of the Cox model is that it makes fewer assumptions about the process underlying the data than do most of the other approaches that can be used with time to event data (Therneau & Grambsch, 2000). In the following section, we will describe the survival and hazard functions, followed by the Cox model. We will then present an example for fitting the model to single-level data to provide the reader with the basic tools needed to fit the Cox model using Mplus, and to serve as an introduction for the multilevel modeling of survival data with the software that we will describe in Chapter 8. In addition, readers who are particularly interested in using these models will have a basic set of tools upon which to build.

Survival and Hazard Functions, and the Cox Model

One of the primary ways in which, time until an event can be characterized is with the survival function. Conceptually, the survival function is simply the probability that an individual will survive at least until time t. It can be written as follows:

$$S(t) = P(T \geq t) \tag{7.8}$$

where T = time when the event occurs.

One useful way in which we can understand time until even data is by plotting the survival function on the *y*-axis and time on the *x*-axis, using a survival curve. We will see an example of a survival curve using Mplus below. The survival function is strictly nonincreasing, meaning that it takes its highest possible value at the beginning of the data collection (typically a value of 1), and then either remains the same (if the event of interest never occurs), or much more likely declines in value as time increases.

Another important function for understanding time to event data is the hazard function, which can be thought of as the instantaneous risk of the event of interest (e.g., death) occurring at a given point in time. Researchers also often examine the cumulative hazard function over time in order to identify how this hazard increases over time. The hazard function is as follows:

$$h(t) = \frac{f(t)}{S(t)} \tag{7.9}$$

where $f(t) = -\dfrac{dS(t)}{dt}$.

Higher values of $h(t)$ at a given time t indicate a higher likelihood of the event occurring.

As noted earlier, one of the most widely used models in survival analysis is based upon the hazard function and is known as the Cox proportional hazards model. This model takes the form as follows:

$$h(t \mid x) = h_0(t)e^{(\beta x)} \tag{7.10}$$

where:
$h(t \mid x)$ is the hazard rate for time t conditional on independent variables, x
x is the set of independent variables
β is the model coefficients linking the independent variables with the conditional hazard rate
$h_0(t)$ is the unspecified baseline hazard function

In many respects, the Cox model is similar to other models that we have examined in this book, in that it relates a set of independent variables, x, to a dependent variable. There are, however, two important differences that need to be noted. First is the fact that the data can be censored, so that for some individuals, the event of interest does not occur before data collection for them is stopped. Thus, when setting up the analysis, we must be sure to include this fact in the model for each individual. Second, a portion of this model is nonparametric, meaning that no assumption is made regarding its underlying distribution. Specifically, no assumption is made about the distributional form of the baseline hazard function, $h_0(t)$. Other models for survival data do make such assumptions, such as $h_0(t)$ follows the exponential or Weibull distributions, for example. However for the Cox model, no assumptions regarding the shape of $h_0(t)$ are made. In other respects, this approach yields results that are similar in form to those we have examined for other models. For example, each independent variable will have associated with it a coefficient expressing the strength and direction of its relationship with the dependent variable, along with a standard error, and associated hypothesis tests. Therefore, just as with linear and logistic regression, the researcher will be able to make statements regarding the statistical significance (or not) of the relationship between each independent variable and the dependent variable, in this case, the conditional hazard rate.

There are two important assumptions that must be satisfied for the Cox model to be appropriately used. First, the censoring of observations must be noninformative. This means that the reasons for an observation to be censored must be unrelated to the probability of the event of interest (e.g., death) occurring. Second, we must assume that the hazard functions for individuals with different levels of the covariates (e.g., members of the treatment and control groups) are proportional over time. Thought another way, when this assumption is met, over time the hazard function for members at one level of the covariate is constant to that of members at another level of the covariate.

Fitting the Cox Model Using Mplus

Fitting survival models using Mplus can be done quite easily. Let us consider data that were described in the second edition of *Applied Survival Analysis: Regression Modeling of Time to Event Data* (Hosmer, Lemeshow, & May, 2008). In this research scenario, researchers have collected data from a clinical trial in which 1,151 AIDS patients were randomly assigned to receive a drug treatment regimen involving Indinavir or a drug treatment not involving Indinavir. The study participants were followed until they either died or had an AIDS-related disease event. The researchers were interested in whether there were differences in the time (measured in days) until death or an AIDS-related event for those who received Indinavir (coded as 0) and those who did not (coded as 1), as well as there existed a relationship between age at enrollment in the study and the time until the event of interest. The following Mplus program fits the Cox proportional hazards model in order to address these research questions.

```
TITLE:     Model 7.12  Survival analysis using the Cox
regression model
DATA:      FILE = single_level_survival_data.csv;
VARIABLE:
    NAMES = time aids trt age;
          SURVIVAL = time;
          TIMECENSORED = aids (0=NOT 1=RIGHT);
analysis:
    BASEHAZARD=OFF;
MODEL:     time ON trt age;
plot:
     type=plot2;
```

Given that much of the program structure is similar to what we have seen
in numerous examples throughout the book already, we will focus only on
those elements that are unique to fitting the survival analysis model. The
SURVIVAL command defines the variable that reflects time until the event
of interest, and TIMECENSORED denotes the variable that indicates whether
the event of interest occurred (death or occurrence of AIDS-related event),
or the individual was censored out of the study. In addition, we can define
the values as 0 (did not occur) or 1 (right censored). Under analysis,
the BASEHAZARD command is used to indicate whether a parametric or
nonparametric model is used, that is, whether the hazard function is
assumed to take a particular form or not. When the OFF option is specified, the nonparametric model is fit to the data. The MODEL statement is
very much like similar statements that we have seen for other models,
where the dependent variable (time) is related to the independent variables (trt and age) using ON. Finally, by specifying plot2, we will be
able to obtain a variety of useful plots which can be used to better understand the relationship of treatment and age with the time until death or
an AIDS-related diagnosis occurs. The relevant output from this command appears as follows:

```
MODEL FIT INFORMATION

Number of Free Parameters                          2

Loglikelihood

          H0 Value                          -741.351
          H0 Scaling Correction Factor        1.0204
             for MLR

Information Criteria

          Akaike (AIC)                      1486.701
          Bayesian (BIC)                    1496.798
```

```
        Sample-Size Adjusted BIC          1490.445
          (n* = (n + 2) / 24)
```

MODEL RESULTS

		Estimate	S.E.	Est./S.E.	Two-Tailed P-Value
TIME	ON				
TRT		-0.685	0.215	-3.185	0.001
AGE		0.020	0.011	1.843	0.065

Mplus provides the user with information criteria, such as the AIC, BIC, and sample size-adjusted BIC, which can be used to compare the fit of several models. Under MODEL RESULTS, the parameter estimates, standard errors, test statistic (Est./S.E.), and associated p-value appear for each variable. From these results, we can see that there was a statistically significant difference in time until the event for the two treatment conditions, such that those who did not receive Indinavir (1) experienced death or an AIDS-related diagnosis sooner than those who received Indinavir. There was no significant relationship between age at entry into the study and the time until the event occurred.

As alluded to earlier, a number of graphical tools are available for the researcher fitting survival models using Mplus. These can be accessed by clicking 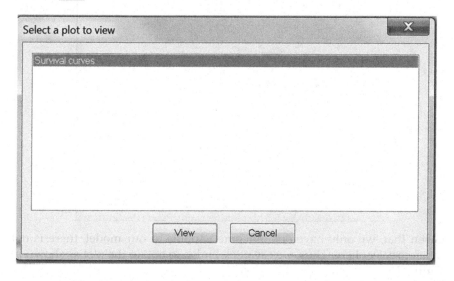 in the menu bar. We then obtain the following window:

By clicking the View button, we then obtain the following window, which we can use to select from a variety of graphing options.

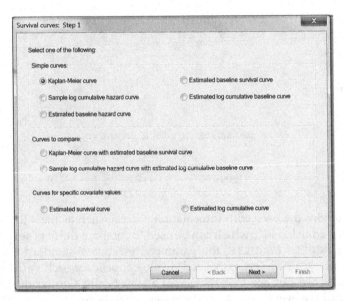

For example, we can view a Kaplan–Meier curve for time until the event, accounting for censoring. This is selected as the first button. We can then click Next to get the following window:

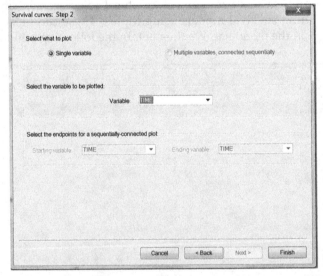

Given that, we only have one outcome variable in our model, there is not much to select here. Thus, we can click Finish to obtain the plot as shown in Figure 7.1.

On the *y*-axis is the probability of survival (i.e., not having the event of interest NOT occur), and on the *x*-axis is time. Thus, at the beginning of the study (day 0), all members of the sample have survived. As time progresses,

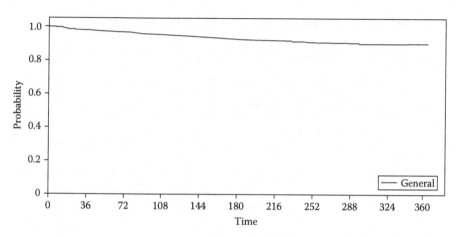

FIGURE 7.1
Survival curve for AIDS-related event or death.

we can see that there is a gradual decline in the survival rate, though it never declines to below 0.9.

In addition to an overall plot of the survival function, it might also be of interest for the researcher to compare survival plots for subgroups within the sample. In the current example, the researcher may want to examine the survival plots for those in the treatment and control conditions separately. Such plots can be obtained with Mplus using the following sequence of commands. First, we must click ⟦⩘⟧, and then select Survival curves from the following window. We would then click on the radio button next to Estimated survival curve under the Curves for specific covariate values: section of the window, as follows:

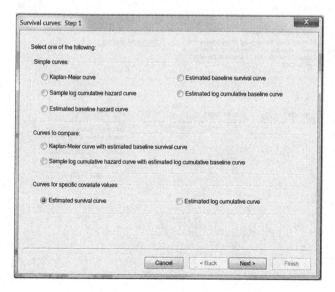

After clicking Next twice, we obtain the following window:

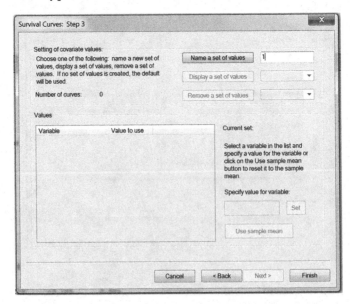

We would then type one of the values of interest in the window next to
| Name a set of values | .

For example, if we were interested in producing separate graphs for the treatment and control conditions, which were coded as 1 and 0 respectively, we would first type 1 into the window.

We would then click Name a set of values after which the window would appear as follows:

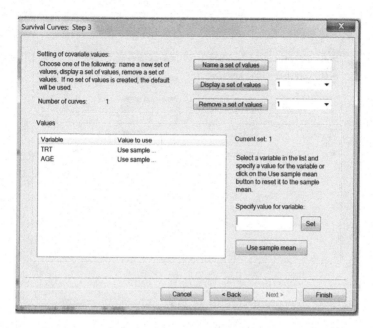

We would then click on TRT (treatment) and type 1 in the box below Specify value for variables: and then click the Set button.

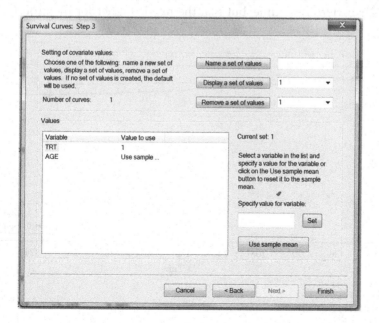

We now have requested a survival curve for the TRT group coded as 1. We would repeat the steps for the group coded as 0.

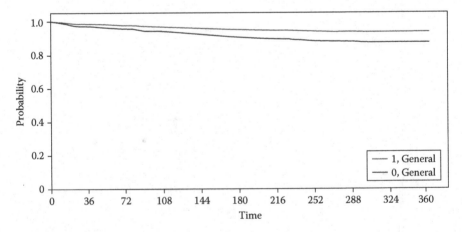

We then click Finish, and the following pair of survival curves will appear (Figure 7.2).

From these results, we can see that the survival rate for each group follows a similar pattern over time, but that the rate is higher for the group coded as 1.

FIGURE 7.2
Survival plots by group.

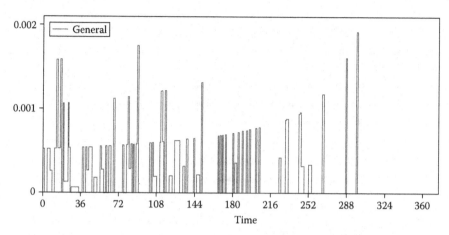

FIGURE 7.3
Hazard rate by time.

It is also possible to plot both the baseline hazard function, and the log of the cumulative hazard function using the graphing facilities in Mplus. In order to plot the baseline hazard function, we would follow the same command sequence as before, and select Estimated baseline hazard curve, after which we click Next and then Finish (Figure 7.3).

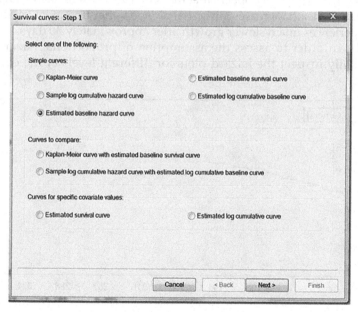

We can see from this plot the periods when the hazard rate is greatest, which appears to occur either early in the study period, or after 220 days. The log of the cumulative hazard function is also a useful graph for understanding time until even data. It can also be obtained using the windows described earlier.

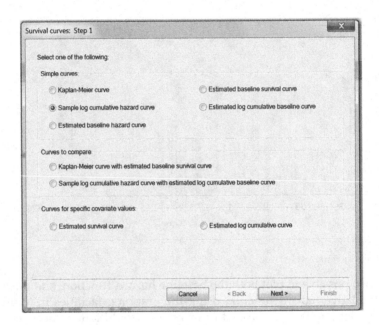

The resulting graph appears as shown in Figure 7.4.

Based on this graph, it appears that the relative hazard of having an AIDS-related event/death increases fairly sharply soon after treatment begins, and then experiences much slower growth after approximately 30 days.

Finally, in order to assess the assumption of proportional hazards, we can visually inspect the hazard plots for different levels of the covariate

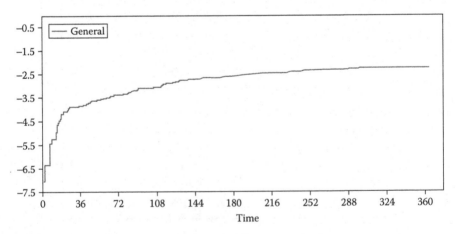

FIGURE 7.4
Log cumulative hazard curve.

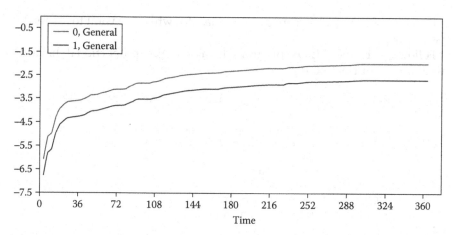

FIGURE 7.5
Log cumulative hazard curve by treatment condition.

and check to see whether they appear to be parallel. As an example, we can use the same sequence of command selections that we employed to obtain separate survival plots for the treatment and control groups, but select Estimated log cumulative curve, rather than Estimated survival curve. In all respects, the command sequence is identical to that described earlier (Figure 7.5).

We are looking to see whether the curves for the two groups appear to be parallel to one another, which does seem to be the case here for times beyond approximately 30 days, but not for the first 30 days.

It is also possible to construct a formal test for the proportional hazards assumption by specifying an interaction between time and the covariate. Essentially, this interaction is testing whether the impact of the covariate is constant over time. If the interaction is statistically significant, then we would conclude that the proportional hazards assumption has indeed been violated, and the nonproportional hazards model should be used. As described in Liu (2012), the researcher using this test needs to make two additional adjustments when calculating this interaction. First, the distribution of time may not be linearly associated with the log of the hazard function, which in turn can lead to estimation problems. Thus, it is recommended that the log of time be used in place of time when calculating the interaction. Second, the covariate and the interaction terms are likely to be highly correlated, thus leading to problems associated with collinearity (parameter estimation instability and inflation of standard errors). Using the centered value of log (time) will generally correct this problem (Liu, 2012). In summary, the proportional hazards assumption can be tested by including in the model the interaction of time and a particular covariate for which we want to assess the assumption.

Following is the Mplus program to assess the proportional hazards assumption for treatment:

```
TITLE: Model 7.13 Survival analysis using the Cox regression
model
     and assessing the proportional hazards assumption
DATA:  FILE = single_level_survival_data.csv;
VARIABLE:
     NAMES = time aids trt age;
         SURVIVAL = time;
         TIMECENSORED = aids (0=NOT 1=RIGHT);
     usevariables time aids trt age int1 ltime;
DEFINE:
     ltime=log(time);
     center ltime (grandmean);
     int1=ltime*trt;
analysis:
     BASEHAZARD=OFF;
MODEL: time ON trt age int1;
plot:
     type=plot2;
```

The key component in this program is the creation of the centered version of the log (time), and the interaction of this new variable with treatment (trt). All of this is done under the DEFINE command, and then the interaction is added to the model. The relevant output appears as follows:

```
THE MODEL ESTIMATION TERMINATED NORMALLY



MODEL FIT INFORMATION

Number of Free Parameters                              5

Loglikelihood

              H0 Value                      -1943.602
              H0 Scaling Correction Factor     2.6404
                 for MLR

Information Criteria

              Akaike (AIC)                   3897.203
              Bayesian (BIC)                 3922.445
              Sample-Size Adjusted BIC       3906.563
                 (n* = (n + 2) / 24)
```

MODEL RESULTS

		Estimate	S.E.	Est./S.E.	Two-Tailed P-Value
TIME	ON				
TRT		-0.978	0.252	-3.874	0.000
AGE		0.025	0.011	2.233	0.026
INT1		-2.466	0.319	-7.727	0.000
Means					
LTIME		0.000	0.022	0.000	1.000
Variances					
LTIME		0.535	0.061	8.796	0.000

The interaction between log (time) and treatment (int1) was statistically significant, indicating that the proportional hazards assumption has been violated, and that the interaction term should remain in the model. Though not included here, a similar analysis was run for age, and the interaction of it with log (time) was also found to be statistically significant, indicating that the proportional hazards assumption for that variable was violated as well.

As noted earlier, there are a number of models that can be used with survival data in the Mplus environment. However, the purpose of this brief discussion was to introduce readers unfamiliar with these models to perhaps the most widely used one, the Cox proportional hazards model. We have shown how this model can be fit using Mplus, and some of the graphical tools available with this software. There are many other models available, and the interested reader is encouraged to Asparouhov (2014) for a description of the variety of survival models that can be fit using Mplus.

Summary

This chapter marks a major change in direction in terms of the type of data upon which we will focus. Through the first six chapters, we have been concerned with models in which the dependent variable is continuous, and generally assumed to be normally distributed. In Chapter 7, we learned about a variety of models designed for categorical dependent variables. In perhaps the simplest instance, such variables can be dichotomous, so that logistic regression is most appropriate for data analysis. When the outcome variable has more than two ordered categories, we saw that logistic regression can be easily extended in the form of the cumulative logits model. We also examined dependent variables that are counts, in which case we may choose

Poisson regression or the negative binomial model, depending upon how frequently the outcome being counted occurs. Finally, we discussed perhaps the most widely used technique for dealing with time until even data, the Cox proportional hazards model. As with Chapter 1, the goal of Chapter 7 was primarily to provide an introduction to the single-level versions of the multilevel models to come. In Chapter 8, we will see that the model types described here can be extended into the multilevel context very much, as standard regression was extended to the multilevel context.

8

Multilevel Generalized Linear Models (MGLMs) and Multilevel Survival Models

In the previous chapter, we introduced generalized linear models, which are useful when the outcome variable of interest is categorical in nature. We described a number of models in this broad family, including logistic regression for binary, ordinal, and multinomial data distributions, as well as Poisson regression models for count or frequency data. In each of those examples, the data were collected at a single level. However, just as is true for normally distributed outcome variables, it is common for categorical variables to be gathered in a multilevel framework. The focus of this chapter is on models designed specifically for scenarios in which the outcome of interest is either categorical or count in nature and the data have been collected in a multilevel framework. Chapter organization will mirror that of Chapter 7, beginning with a description of fitting logistic regression for dichotomous data, followed by models for ordinal and nominal dependent variables, and concluding with models for count data that fit the Poisson distribution, and for the case of overdispersed counts. Given that the previous chapter provided the relevant mathematical underpinnings for these various models in the single-level case and Chapter 2 introduced some of the theory underlying multilevel models, the current chapter will focus almost exclusively on the application of Mplus to fit these models and on the interpretation of the resultant output.

Multilevel Generalized Linear Models (MGLMs) for a Dichotomous Outcome Variable

In order to introduce MGLMs for dichotomous outcomes, let us consider the following example. A researcher has collected testing data that indicate whether 9316 students have passed a state mathematics assessment, along with several measures of mathematics aptitude that were measured before administration of the achievement test. She is interested in whether a relationship exists between the score in number sense aptitude and the likelihood that a student will achieve a passing score in the mathematics achievement test, for which all examinees are categorized as either passing (1) or failing (0). Given that the outcome variable is dichotomous, we could use the binary logistic regression method introduced in Chapter 7. However, students in this sample are clustered by school, as was the case with the data that were examined in Chapters 3 and 4. Therefore, we will need to appropriately account for this multilevel data structure in our regression analysis.

Random Intercept Logistic Regression

The Mplus program needed to fit the multilevel dichotomous logistic regression model represents a mixture of examples that we saw in Chapters 4 and 7. Note that in this initial analysis, we have a fixed effect for the intercept and the slope of the independent variable numsense, but we allow only a random intercept, thereby assuming that the relationship between the number sense score and the likelihood of achieving a passing score in the state mathematics assessment (score2) is fixed across schools; that is, the relationship of numsense with score2 does not vary from one school to another. Remember from earlier chapters that the default setting with Mplus is that intercepts are random effects. Many of the modeling commands for fitting multilevel logistic regression are identical to those for single-level logistic regression models, and therefore, these will not be discussed again here. The clustering variable, school, is explicitly identified in the cluster = statement, and level 1 covariates are identified in the within = statement. In the ANALYSIS section of the program, we identify the model as being a two-level multilevel, and then, in the MODEL section, the level 1 (%within%) and level 2 (%between%) components of the model are defined. In this example, we have only a single level 1 covariate (numsense) and nothing for level 2. In the OUTPUT line, we request tech1 and tech8 outputs, which provide us with the parameter specifications and the optimization history of the model-fitting algorithm. In addition, we also request standardized estimates, using the standardized subcommand.

```
TITLE:   Model 8.1 Multilevel dichotomous logistic regression
with random intercept;

DATA:    file is mathfinal.csv;

VARIABLE:
names=numsense computation L_free female score score2 school;
missing=.;
usevariables=numsense score2 school;
categorical=score2;
cluster=school;
within=numsense;
```

```
ANALYSIS:
type=twolevel;

MODEL:
%within%
score2 on numsense;

%between%

OUTPUT:
tech1 tech8;
```

The resulting output appears below.

```
SUMMARY OF DATA

    Number of missing data patterns          1
    Number of y missing data patterns        0
    Number of u missing data patterns        1
    Number of clusters                       40
```

```
COVARIANCE COVERAGE OF DATA

Minimum covariance coverage value    0.100

UNIVARIATE PROPORTIONS AND COUNTS FOR CATEGORICAL VARIABLES

    SCORE2
        Category 1     0.435          4049.000
        Category 2     0.565          5267.000
```

```
THE MODEL ESTIMATION TERMINATED NORMALLY
```

```
MODEL FIT INFORMATION

Number of Free Parameters                    3

Loglikelihood

        H0 Value                        -4914.940
        H0 Scaling Correction Factor       2.1998
            for MLR
```

Information Criteria

        Akaike (AIC)   9835.880
        Bayesian (BIC)   9857.298
        Sample-Size Adjusted BIC   9847.765
           (n* = (n + 2) / 24)

MODEL RESULTS

	Estimate	S.E.	Est./S.E.	Two-Tailed P-Value
Within Level				
SCORE2 ON				
NUMSENSE	0.059	0.003	19.649	0.000
Between Level				
Thresholds				
SCORE2$1	-0.386	0.089	-4.356	0.000
Variances				
SCORE2	0.286	0.085	3.365	0.001

LOGISTIC REGRESSION ODDS RATIO RESULTS

Within Level

    SCORE2   ON
        NUMSENSE   1.061

The output that we obtain from this analysis is very similar in format to the output for the single-level model that we worked with in Chapter 7. We first examine the variability in intercepts from school to school, which is presented as the between-level variance of SCORE2, or 0.286. The mean value of the intercept across schools is −0.386. With regard to the fixed effect, the slope of numsense, we see that higher scores are associated with a greater likelihood of passing the state mathematics assessment, with the slope being 0.059 ($p < 0.05$). Remember that for this analysis, the larger value of the outcome in the numerator of the logit, in this case, passing, was coded as 1, whereas failing was coded as 0. The standard error, test statistic, and p-value appear in the next three columns. The results are statistically significant ($p < 0.001$), leading to the conclusion that overall, number sense scores are positively related to the likelihood of a student achieving a passing score in the assessment.

Random Coefficient Logistic Regression

As with the linear multilevel models, it is also possible to allow for random slopes with multilevel generalized linear models (GLiMs). The command structure for Mplus is very similar to that used above, with a few adjustments under the ANALYSIS: and MODEL: commands, as can be seen below. In all other respects, the program for model 8.2 is very similar to that for model 8.1. Two primary differences are evident in the program for fitting model 8.2. First, the keyword random is added in the type = subcommand under ANALYSIS. Second, we must explicitly define the random slope effect in the MODEL command. In Mplus, the random slope is a variable, which we have named s and which is defined in the line s | score2 on numsense;.

```
TITLE:  Model 8.2  Multilevel dichotomous logistic regression
with random intercept and random slope;

DATA:   file is mathfinal.csv;

VARIABLE:
names=numsense computation L_free female score score2 school;
missing=.;
usevariables=numsense score2 school;
categorical=score2;
cluster=school;
within=numsense;

DEFINE:
center numsense (grandmean);

ANALYSIS:
type=twolevel random;

MODEL:
%within%
s | score2 on numsense;

%between%
score2 with s;

OUTPUT:
tech1 tech8 standardized;
savedata:
results are model82.out;
```

The output from this model appears in the following. Note that owing to space considerations, we report only the results for the case when both the dependent and independent variables are standardized (STDYX); however, Mplus also reports results for the case when only the dependent and only the independent variable, respectively, are standardized. However, in most situations, we will be interested in the results for the case when both variables are standardized.

```
SUMMARY OF DATA

    Number of missing data patterns              1
    Number of y missing data patterns           0
    Number of u missing data patterns           1
    Number of clusters                          40



COVARIANCE COVERAGE OF DATA

Minimum covariance coverage value     0.100


UNIVARIATE PROPORTIONS AND COUNTS FOR CATEGORICAL VARIABLES

    SCORE2
        Category 1     0.435         4049.000
        Category 2     0.565         5267.000



THE MODEL ESTIMATION TERMINATED NORMALLY



MODEL FIT INFORMATION

Number of Free Parameters                       5

Loglikelihood

        H0 Value                         -4879.275
        H0 Scaling Correction Factor        1.2716
            for MLR

Information Criteria

        Akaike (AIC)                     9768.551
        Bayesian (BIC)                   9804.248
        Sample-Size Adjusted BIC         9788.359
            (n* = (n + 2) / 24)
```

MODEL RESULTS

	Estimate	S.E.	Est./S.E.	Two-Tailed P-Value
Within Level				
Between Level				
SCORE2 WITH				
S	-0.011	0.006	-1.719	0.086
Means				
S	0.065	0.005	14.055	0.000
Thresholds				
SCORE2$1	-0.287	0.130	-2.210	0.027
Variances				
SCORE2	0.454	0.171	2.662	0.008
S	0.000	0.000	1.322	0.186

STANDARDIZED MODEL RESULTS

STDYX Standardization

	Estimate	S.E.	Est./S.E.	Two-Tailed P-Value
Within Level				
Between Level				
SCORE2 WITH				
S	-0.882	0.082	-10.750	0.000
Means				
S	3.679	1.182	3.113	0.002
Thresholds				
SCORE2$1	-0.426	0.252	-1.692	0.091
Variances				
SCORE2	1.000	0.000	999.000	999.000
S	1.000	0.000	999.000	999.000

We will focus on aspects of the output for the random coefficient model output that differ from that of the random intercepts. In particular, note that we have an estimate of the variance of the U_{1j} estimates for specific schools. This value, which is less than 0.001, is relatively small when compared with the variation of intercepts across schools and of individuals within schools, meaning that the relationship of number sense with the likelihood of an individual receiving a passing score in the mathematics achievement test is relatively similar across the schools. Note that, although Mplus output yields numeric results to only three significant digits, the user can obtain more detailed estimates by using the savedata command, as demonstrated above. The parameter estimates and standard errors for the raw and standardized results are saved in the file model82.out. By referring to these results, we see that the variance of U_{ij} is 0.0003. The modal slope across schools is 0.065, again indicating that individuals with higher number sense scores also have a higher likelihood of passing the mathematics assessment. Finally, it is important to note that the correlation between the random components of the slope and intercept, −0.882, the standardized version of τ_{01}, is very strongly negative. This value appears in the STDYX section of the output.

Inclusion of Additional Level 1 and Level 2 Effects to the Multilevel Logistic Regression Model

The researcher in our example is also interested in learning whether there is a statistically significant relationship between an examinee's gender (female, where 1 = female, and 0 = male) and the likelihood of passing the state mathematics assessment, as well as the relationship between passing and number sense score. In order to fit the additional level 1 variable to the random coefficient model, we would use the following Mplus program to obtain the subsequent output. This program fits a model in which the impacts of both the number sense score and an examinee's gender are allowed to vary across schools; that is, two random slopes are present.

```
TITLE:   Model 8.3  Multilevel dichotomous logistic regression
with random intercept and random slope for two level 1
covariates;

DATA:    file is mathfinal.csv;

VARIABLE:
names=numsense computation L_free female score score2 school;
missing=.;
usevariables=numsense score2 female school;
categorical=score2;
cluster=school;
within=numsense female;
```

```
DEFINE:
center numsense (grandmean);

ANALYSIS:
type=twolevel random;

MODEL:
%within%
s1 | score2 on numsense;
s2 | score2 on female;

%between%
score2 with s1;
score2 with s2;

OUTPUT:
tech1 tech8 standardized;

savedata:
results are model83.out;
```

SUMMARY OF DATA

Number of missing data patterns	1
Number of y missing data patterns	0
Number of u missing data patterns	1
Number of clusters	40

COVARIANCE COVERAGE OF DATA

Minimum covariance coverage value   0.100

UNIVARIATE PROPORTIONS AND COUNTS FOR CATEGORICAL VARIABLES

SCORE2
| Category 1 | 0.434 | 4044.000 |
| Category 2 | 0.566 | 5267.000 |

THE MODEL ESTIMATION TERMINATED NORMALLY

MODEL FIT INFORMATION

Number of Free Parameters   8

Loglikelihood

      H0 Value   -4876.639
      H0 Scaling Correction Factor   1.2011
         for MLR

Information Criteria

      Akaike (AIC)   9769.279
      Bayesian (BIC)   9826.391
      Sample-Size Adjusted BIC   9800.968
        (n* = (n + 2) / 24)

MODEL RESULTS

	Estimate	S.E.	Est./S.E.	Two-Tailed P-Value
Within Level				
Between Level				
SCORE2 WITH				
S1	-0.010	0.006	-1.628	0.104
S2	0.021	0.039	0.533	0.594
Means				
S1	0.065	0.004	14.545	0.000
S2	0.044	0.056	0.788	0.431
Thresholds				
SCORE2$1	-0.283	0.124	-2.281	0.023
Variances				
SCORE2	0.414	0.179	2.314	0.021
S1	0.000	0.000	1.719	0.086
S2	0.005	0.024	0.201	0.841

STANDARDIZED MODEL RESULTS

STDYX Standardization

	Estimate	S.E.	Est./S.E.	Two-Tailed P-Value

```
Within Level

Between Level

    SCORE2    WITH
      S1              -0.768       0.124      -6.184       0.000
      S2               0.468       1.769       0.265       0.791

    Means
      S1               3.207       0.761       4.214       0.000
      S2               0.638       1.673       0.382       0.703

    Thresholds
      SCORE2$1        -0.439       0.258      -1.704       0.088

    Variances
      SCORE2           1.000       0.000     999.000     999.000
      S1               1.000       0.000     999.000     999.000
      S2               1.000       0.000     999.000     999.000
```

These results indicate that being female is not significantly related to one's likelihood of passing the mathematics achievement test; that is, there are no gender differences in the likelihood of passing. In addition, there were no statistically significant differences across schools in the relationship of gender and the likelihood of passing, as evidenced by the hypothesis test for the variance of the random component of slopes (0.005), with a p-value of 0.0841. The average slope for gender across schools was 0.021, but it was not statistically significant, meaning that there were no differences in the likelihood of passing between males and females.

We can also include level 2 independent variables in the model, such as the proportion of students receiving free lunch at each school (L_Free). In Model 8.4, we fit a random intercept model, including the level 2 independent variable L_Free. The random coefficient terms have been removed from this example for the sake of simplicity.

```
TITLE:   Model 8.4  Multilevel dichotomous logistic regression
with random intercept, two level 1 covariates, and one level 2
covariate;

DATA:    file is mathfinal.csv;

VARIABLE:
names=numsense computation L_free female score score2 school;
missing=.;
usevariables=numsense score2 female school L_free;
categorical=score2;
cluster=school;
within=numsense female;
```

```
between=L_free;

DEFINE:
center numsense (grandmean);

ANALYSIS:
type=twolevel;

MODEL:
%within%
score2 on numsense female;

%between%
score2 on L_free;

OUTPUT:
tech1 tech8 standardized;

savedata:
results are model84.out;
```

SUMMARY OF DATA

Number of missing data patterns	1
Number of y missing data patterns	0
Number of u missing data patterns	1
Number of clusters	34

COVARIANCE COVERAGE OF DATA

Minimum covariance coverage value   0.100

UNIVARIATE PROPORTIONS AND COUNTS FOR CATEGORICAL VARIABLES

SCORE2

Category 1	0.432	3052.000
Category 2	0.568	4014.000

THE MODEL ESTIMATION TERMINATED NORMALLY

MODEL FIT INFORMATION

Number of Free Parameters   5

```
Loglikelihood

        H0 Value                            -3644.908
        H0 Scaling Correction Factor          1.2624
          for MLR

Information Criteria
        Akaike (AIC)                         7299.815
        Bayesian (BIC)                       7334.131
        Sample-Size Adjusted BIC             7318.242
          (n* = (n + 2) / 24)
```

MODEL RESULTS

	Estimate	S.E.	Est./S.E.	Two-Tailed P-Value
Within Level				
SCORE2 ON				
NUMSENSE	0.064	0.003	22.901	0.000
FEMALE	-0.026	0.057	-0.455	0.649
Between Level				
SCORE2 ON				
L_FREE	-0.009	0.003	-2.657	0.008
Thresholds				
SCORE2$1	-0.912	0.210	-4.351	0.000
Residual Variances				
SCORE2	0.279	0.082	3.395	0.001

LOGISTIC REGRESSION ODDS RATIO RESULTS

Within Level

```
   SCORE2      ON
     NUMSENSE        1.066
     FEMALE          0.974
```

```
STANDARDIZED MODEL RESULTS


STDYX Standardization

                                                           Two-Tailed
                      Estimate        S.E.   Est./S.E.      P-Value

Within Level

   SCORE2      ON
      NUMSENSE            0.639       0.017      38.100        0.000
      FEMALE            -0.006       0.012      -0.456        0.648

Between Level

   SCORE2      ON
      L_FREE            -0.422       0.130      -3.254        0.001

   Thresholds
      SCORE2$1          -1.565       0.320      -4.899        0.000

   Residual Variances
      SCORE2             0.822       0.110       7.503        0.000
```

There is a statistically significant negative relationship between the proportion of students in the school receiving free lunch and the likelihood of an individual student passing the mathematics assessment, with a coefficient value of −0.009 and a *p*-value of 0.008. Results for the other predictor variables are largely the same as in the prior analyses.

It is possible for us to compare the fit of the various models in order to determine which is optimal, based on statistical fit. This can be done by using the information criteria, Akaike's information criterion (AIC), Bayesian information criterion (BIC), and sample size adjusted BIC (aBIC). As we have noted in previous chapters, the models with the smaller value of these index values are said to fit the data better. The fit indices for Models 8.1 through 8.4 appear in Table 8.1, below.

TABLE 8.1

Information Indices for Models 8.1 through 8.4

Model	AIC	BIC	aBIC
8.1	98355.880	9857.298	9847.765
8.2	9768.551	9804.248	9788.359
8.3	9769.279	9826.391	9800.968
8.4	7299.815	7334.131	7318.242

From these results, it is clear that of the models that were investigated here, Model 8.4 provides the best fit to the data. Thus, including the percentage of students on free lunch at each school led to markedly improved fit of the model to the data.

MGLM for an Ordinal Outcome Variable

As was the case for non-multilevel data, the cumulative logits link function can be used with ordinal data in the context of multilevel logistic regression. Indeed, the link will be the familiar cumulative logit that we described in Chapter 7. Furthermore, the multilevel aspects of the model, including random intercept and coefficient, take the same form as the one we described above. To provide context, let us again consider the mathematics achievement results for students. In this case, the outcome variable takes one of the three possible values for each member of the sample: 1 = Failure, 2 = Pass, 3 = Pass with distinction. In this case, the question of most interest to the researcher is whether a computation aptitude score is a good predictor of status on the mathematics achievement test.

Random Intercept Ordinal Logistic Regression

In order to fit a multilevel cumulative logits model using Mplus, we will use programs that are very similar to those described above. In addition, model parameter estimation is achieved using maximum likelihood, based on the Newton–Raphson method. In the first example, we will examine the relationship between computation scores in a formative exam and the ordinal scores in the final mathematics test. The Mplus program to fit this model appears below, followed by the output.

```
TITLE:   Model 8.5 Multilevel ordinal logistic regression with
random intercept and one level 1 covariate;

DATA:    file is mathfinal.csv;

VARIABLE:
names=numsense computation L_free female score score2 school;
missing=.;
usevariables=computation score school;
categorical=score;
cluster=school;
within=computation;
```

```
DEFINE:
center computation (grandmean);

ANALYSIS:
type=twolevel;

MODEL:
%within%
score on computation;

%between%

OUTPUT:
tech1 tech8 standardized;

savedata:
results are model85.out;
```

SUMMARY OF DATA

```
    Number of missing data patterns           1
    Number of y missing data patterns         0
    Number of u missing data patterns         1
    Number of clusters                       40
```

COVARIANCE COVERAGE OF DATA

Minimum covariance coverage value   0.100

UNIVARIATE PROPORTIONS AND COUNTS FOR CATEGORICAL VARIABLES

```
    SCORE
      Category 1    0.435        4049.000
      Category 2    0.480        4468.000
      Category 3    0.086         799.000
```

THE MODEL ESTIMATION TERMINATED NORMALLY

MODEL FIT INFORMATION

Number of Free Parameters   4

Loglikelihood

      H0 Value   -7175.050
      H0 Scaling Correction Factor   2.7468
        for MLR

Information Criteria

      Akaike (AIC)   14358.101
      Bayesian (BIC)   14386.659
      Sample-Size Adjusted BIC   14373.948
        (n* = (n + 2) / 24)

MODEL RESULTS

	Estimate	S.E.	Est./S.E.	Two-Tailed P-Value
Within Level				
SCORE ON				
COMPUTATION	0.050	0.003	17.992	0.000
Between Level				
Thresholds				
SCORE$1	-0.415	0.091	-4.549	0.000
SCORE$2	2.921	0.145	20.183	0.000
Variances				
SCORE	0.362	0.106	3.428	0.001

LOGISTIC REGRESSION ODDS RATIO RESULTS

Within Level

  SCORE   ON
    COMPUTATION   1.051

BRANT WALD TEST FOR PROPORTIONAL ODDS

	Chi-Square	Degrees of Freedom	P-Value
SCORE			
Overall test	0.149	1	0.700
COMPUTATION	0.149	1	0.700

STANDARDIZED MODEL RESULTS

STDYX Standardization

	Estimate	S.E.	Est./S.E.	Two-Tailed P-Value
Within Level				
SCORE ON				
COMPUTATION	0.539	0.021	25.149	0.000
Between Level				
Thresholds				
SCORE$1	-0.690	0.167	-4.122	0.000
SCORE$2	4.857	0.764	6.354	0.000
Variances				
SCORE	1.000	0.000	999.000	999.000

An examination of the results presented above reveals that the variance of intercepts across schools is 0.362 and that this term is statistically significant (p = 0.001). Therefore, we would conclude that there are differences in intercepts from one school to the next. In addition, we see that there is a significant positive relationship between performance on the computation aptitude subtest and performance on the mathematics achievement test, indicating that examinees who have higher computation skills also are more likely to attain higher ordinal scores in the achievement test, for example, pass versus fail and pass with distinction versus pass. We also obtain estimates of the average model intercepts across schools, which are termed thresholds here, as well as a test for the proportional odds assumption, which we discussed in Chapter 7. The lack of statistical significance for this test (p = 0.700) means that this assumption was not violated and that the single coefficient estimate for computation (0.050) applies to all levels of the dependent variable. Finally, we also have available the AIC, BIC, and aBIC values, which we can use to compare the relative fit of this to other models.

Random Intercept Ordinal Logistic Regression with Predictors at Levels 1 and 2

As an example of fitting models with both level 1 and level 2 variables, let us include the proportion of students receiving free lunch in the schools (L_Free) as an independent variable along with the computation score.

```
TITLE:   Model 8.6  Multilevel ordinal logistic regression with
random intercept, one level 1 covariate, and one level 2
covariate;

DATA:    file is mathfinal.csv;

VARIABLE:
names=numsense computation L_free female score score2 school;
missing=.;
usevariables=computation score school L_free;
categorical=score;
cluster=school;
within=computation;
between=L_free;

DEFINE:
center computation (grandmean);

ANALYSIS:
type=twolevel;

MODEL:
%within%
score on computation;

%between%
score on L_free;

OUTPUT:
tech1 tech8 standardized;

savedata:
results are model86.out;
```

SUMMARY OF DATA

Number of missing data patterns	1
Number of y missing data patterns	0
Number of u missing data patterns	1
Number of clusters	34

COVARIANCE COVERAGE OF DATA

Minimum covariance coverage value   0.100

```
UNIVARIATE PROPORTIONS AND COUNTS FOR CATEGORICAL VARIABLES

   SCORE
     Category 1      0.432                      3055.000
     Category 2      0.476                      3364.000
     Category 3      0.092                       650.000


THE MODEL ESTIMATION TERMINATED NORMALLY



MODEL FIT INFORMATION

Number of Free Parameters                              5

Loglikelihood

         H0 Value                           -5442.560
         H0 Scaling Correction Factor          1.7597
            for MLR

Information Criteria

         Akaike (AIC)                       10895.120
         Bayesian (BIC)                     10929.437
         Sample-Size Adjusted BIC           10913.548
            (n* = (n + 2) / 24)

MODEL RESULTS

                                                      Two-Tailed
                      Estimate    S.E.   Est./S.E.    P-Value

Within Level

   SCORE      ON
      COMPUTATION       0.053    0.003     19.864       0.000

Between Level

   SCORE      ON
      L_FREE           -0.011    0.003     -3.478       0.001

   Thresholds
      SCORE$1          -1.056    0.211     -5.007       0.000
      SCORE$2           2.260    0.246      9.183       0.000

   Residual Variances
      SCORE             0.293    0.086      3.413       0.001
```

LOGISTIC REGRESSION ODDS RATIO RESULTS

Within Level

```
    SCORE        ON
      COMPUTATION         1.054
```

BRANT WALD TEST FOR PROPORTIONAL ODDS

	Chi-Square	Degrees of Freedom	P-Value
SCORE			
Overall test	18.803	2	0.000
COMPUTATION	2.638	1	0.104
L_FREE	16.737	1	0.000

STANDARDIZED MODEL RESULTS

STDYX Standardization

	Estimate	S.E.	Est./S.E.	Two-Tailed P-Value
Within Level				
SCORE ON				
COMPUTATION	0.565	0.020	28.776	0.000
Between Level				
SCORE ON				
L_FREE	-0.509	0.105	-4.839	0.000
Thresholds				
SCORE$1	-1.678	0.259	-6.477	0.000
SCORE$2	3.591	0.844	4.253	0.000
Residual Variances				
SCORE	0.741	0.107	6.917	0.000

Given that we have already discussed the results of the previous model in some detail, we will not reiterate those basic ideas again. However, it is important to note those aspects that are different here. Specifically, the variability in the intercepts declined from 0.362 to 0.293 with the inclusion of the school level variable, L_Free. In addition, we see that there is a significant negative relationship between the proportion of individuals receiving free lunch in the school and the likelihood that an individual student will obtain a higher achievement test score. Finally, a comparison of the information indices for the computation-only model and the computation-and-free-lunch model clearly shows that the latter provides a better fit to the data, given its smaller AIC, BIC, and aBIC values. Therefore, we can conclude that including both free lunch and computation score when modeling the three-level achievement outcome variable is preferable to including only computation.

MGLM for Count Data

In the previous chapter, we examined statistical models designed for use with outcome variables that represented the frequency of some event occurring. Typically, these events were relatively rare, such as the number of babies in a family. Perhaps, the most common distribution associated with such counts is the Poisson, a distribution in which the mean and the variance are equal. However, as we saw in Chapter 7, this equality of the two moments does not always hold in all empirical contexts, in which case we have what is commonly referred to as overdispersed data. In such cases, the Poisson regression model relating one or more independent variables to a count-dependent variable is not appropriate and we must make use of the negative binomial distribution, which is able to appropriately model the inequality of the mean and the variance. It is a fairly straightforward matter to extend either of these models to the multilevel context, both conceptually and by using Mplus. In the following sections, we will demonstrate analysis of multilevel count data outcomes in the context of Poisson regression and negative binomial regression by using Mplus. The example to be used involves the number of cardiac warning incidents (e.g., chest pain, shortness of breath, and dizzy spells) for 1000 patients associated with 110 cardiac rehabilitation facilities in a large state over a 6-month period. Patients who had recently suffered from a heart attack and who were entering rehabilitation agreed to be randomly assigned to either a new exercise treatment program or the standard treatment protocol. Of particular interest to the researcher heading up this study is the relationship between treatment condition and the number of cardiac warning incidents. The new approach to rehabilitation is expected to result in fewer such incidents as compared with the traditional method. In addition, the researcher has also collected data on the sex of the patients and the number of hours for which each rehabilitation facility is open during the week. This latter variable is of interest as it reflects the overall availability of the rehabilitation programs. The new method of conducting cardiac rehabilitation is coded in the data as 1, while the standard approach is coded as 0. Males are also coded as 1, while females are assigned a value of 0.

Random Intercept Poisson Regression

The Mplus program and the resultant output for fitting the multilevel Poisson regression model to the data appear below.

```
TITLE: Model 8.7  Multilevel Poisson regression model
DATA:  FILE IS rehab.csv;
VARIABLE:    NAMES ARE heart trt sex hours rehab;
             count IS heart;
      usevariables are heart trt rehab sex;
      cluster=rehab;
      within=trt sex;
ANALYSIS:
      type=twolevel;
MODEL:
      %within%
    heart ON trt sex;

      %between%
      heart;

OUTPUT:
      tech1 tech8 standardized;

savedata:
results are model87.out;

SUMMARY OF DATA

   Number of clusters                          110


COUNT PROPORTION OF ZERO, MINIMUM AND MAXIMUM VALUES

   HEART        0.590          0        225


   WARNING:  COUNT VARIABLE HAS LARGE VALUES.
   IT MAY BE MORE APPROPRIATE TO TREAT SUCH VARIABLES AS
CONTINUOUS.


THE MODEL ESTIMATION TERMINATED NORMALLY


MODEL FIT INFORMATION

Number of Free Parameters                       4
```

Loglikelihood

      H0 Value   -5213.398
      H0 Scaling Correction Factor   13.3422
        for MLR

Information Criteria

      Akaike (AIC)   10434.795
      Bayesian (BIC)   10454.426
      Sample-Size Adjusted BIC   10441.722
        (n* = (n + 2) / 24)

MODEL RESULTS

	Estimate	S.E.	Est./S.E.	Two-Tailed P-Value
Within Level				
HEART ON				
TRT	-0.535	0.192	-2.785	0.005
SEX	0.420	0.162	2.589	0.010
Between Level				
Means				
HEART	0.875	0.148	5.919	0.000
Variances				
HEART	1.493	0.260	5.739	0.000

STANDARDIZED MODEL RESULTS

STDYX Standardization

	Estimate	S.E.	Est./S.E.	Two-Tailed P-Value
Within Level				
HEART ON				
TRT	-0.803	0.128	-6.295	0.000
SEX	0.630	0.166	3.800	0.000

Between Level

  Means
    HEART   0.716   0.140   5.118   0.000

  Variances
    HEART   1.000   0.000   999.000   999.000

In terms of the actual programming, the syntax for Model 8.7 is very similar to that used for the dichotomous logistic regression model. The dependent and independent variables are linked in the usual way that we have seen in Mplus. Here, the outcome variable is heart, which reflects the frequency of the warning signs for heart problems that we described above. The independent variables are treatment (trt) and sex of the individual, while the specific rehabilitation facility is contained in the variable rehab. In this model, we are fitting only a random intercept, with no random slope and no rehabilitation center level variables.

When reviewing the results, we should first take note of the warning that Mplus provides regarding the large values of the dependent variable for some members of the sample. This may be an indication that we will want to try the negative binomial model. The results of the analysis for the Poisson regression model indicate that there is statistically significant variation among the intercepts from rehabilitation facility to rehabilitation facility, with a variance estimate of 1.493. As a reminder, the intercept reflects the mean frequency of events when the independent variables are 0, that is, females in the control condition. The average intercept across the 110 rehabilitation centers is 0.875. Put another way, we can conclude that the mean number of cardiac warning signs varies across rehabilitation centers and that the average female in the control condition will have approximately 0.875 such incidents over the course of 6 months. In addition, these results reveal a negative relationship between heart and trt and a significant positive relationship between heart and sex. Remember that the new treatment is coded as 1 and the control is coded as 0, so that a negative relationship indicates that there are fewer warning signs over 6 months for those in the treatment group than for those in the control group. In addition, males were coded as 1 and females were coded as 0, so that the positive slope for sex means that males have more warning signs on average than females.

Random Coefficient Poisson Regression

If we believe that the treatment will have different impacts on the number of warning signs present among the rehabilitation centers, we would want to fit the random coefficient model. This can be done for Poisson regression, just as it was syntactically for dichotomous logistic regression, as demonstrated in Model 8.8.

```
TITLE: Model 8.8  Multilevel Poisson regression model
DATA:  FILE IS rehab.csv;
VARIABLE:    NAMES ARE heart trt sex hours rehab;
             count IS heart;
       usevariables are heart trt rehab sex;
       cluster=rehab;
       within=trt sex;

ANALYSIS:
       type=twolevel random;
MODEL:
       %within%
       heart ON sex;
       s1 | heart ON trt;

       %between%
       heart;
       s1 with heart;

OUTPUT:
       tech1 tech8;

savedata:
results are model88.out;

SUMMARY OF DATA

   Number of clusters                          110



COUNT PROPORTION OF ZERO, MINIMUM AND MAXIMUM VALUES

   HEART          0.590          0          225



       WARNING:  COUNT VARIABLE HAS LARGE VALUES.
       IT MAY BE MORE APPROPRIATE TO TREAT SUCH VARIABLES AS
       CONTINUOUS.
THE MODEL ESTIMATION TERMINATED NORMALLY



MODEL FIT INFORMATION

Number of Free Parameters                       6
```

```
Loglikelihood

          H0 Value                           -4519.919
          H0 Scaling Correction Factor          5.3206
              for MLR

Information Criteria

          Akaike (AIC)                        9051.838
          Bayesian (BIC)                      9081.284
          Sample-Size Adjusted BIC            9062.228
              (n* (n + 2) / 24)
```

```
MODEL RESULTS

                                                       Two-Tailed
                        Estimate      S.E.   Est./S.E.  P-Value

Within Level

   HEART        ON
      SEX          0.426       0.194      2.191       0.028

Between Level

   S1          WITH
      HEART       -1.744       0.502     -3.471       0.001

   Means
   S1          -0.064       0.131     -0.487       0.626
      HEART        0.394       0.166      2.383       0.017

   Variances
   S1           2.438       0.589      4.135       0.000
      HEART        2.547       0.520      4.898
```

The program for the inclusion of random slopes in the model is very similar to that used with logistic regression for a random coefficient model with one exception. For count data models, standardized coefficients are not available by using Mplus. The random effect for slopes across rehabilitation centers was estimated as 2.438 and was statistically significant, indicating that there is some differential center impact on the number of cardiac warning signs. Similarly, the random intercept variance term was also significant (2.547), as was true for Model 8.7. The covariance of the random slope and intercept model components is also very small (0.082). The average slope for treatment across centers (−0.064) was no longer statistically significant but remained negative. In addition, the relationship between the random effect for the coefficient and the intercept was significantly negative (−1.744), meaning that centers with a larger average number of incidents exhibited a smaller relationship between the treatment effect and the number of incidents.

Inclusion of Additional Level 2 Effects to the Multilevel Poisson Regression Model

You may recall that in addition to testing for treatment and gender differences in the rate of heart warning signs, the researcher conducting this study also wanted to know whether the number of hours per week for which the rehabilitation centers were open (hours) was related to the outcome variable. In order to address this question, we will need to fit a model with both level 1 (trt and sex) and level 2 (hours) effects. Note that for the within-cluster variables, we fit only random intercepts and not the random coefficients. This choice was made only for simplicity sake, and these terms could have been included.

```
TITLE: Model 8.9  Multilevel Poisson regression model
DATA:  FILE IS rehab.csv;
VARIABLE:    NAMES ARE heart trt sex hours rehab;
             count IS heart;
       usevariables are heart trt rehab sex hours;
       cluster=rehab;
       within=trt sex;
       between=hours;

ANALYSIS:
       type=twolevel;
MODEL:
       %within%
     heart ON sex;
       heart ON trt;

       %between%
       heart;
       heart ON hours;

OUTPUT:
       tech1 tech8;

savedata:
results are model89.out;

SUMMARY OF DATA

   Number of clusters                       110
```

```
COUNT PROPORTION OF ZERO, MINIMUM AND MAXIMUM VALUES

    HEART          0.590           0        225
```

```
    WARNING:  COUNT VARIABLE HAS LARGE VALUES.
    IT MAY BE MORE APPROPRIATE TO TREAT SUCH VARIABLES AS
CONTINUOUS.
```

```
THE MODEL ESTIMATION TERMINATED NORMALLY
```

```
MODEL FIT INFORMATION
Number of Free Parameters                        5

Loglikelihood

            H0 Value                      -5210.592
            H0 Scaling Correction Factor     10.8779
               for MLR

Information Criteria

            Akaike (AIC)                   10431.183
            Bayesian (BIC)                 10455.722
            Sample-Size Adjusted BIC       10439.842
               (n* = (n + 2) / 24)
```

```
MODEL RESULTS
```

	Estimate	S.E.	Est./S.E.	Two-Tailed P-Value
Within Level				
HEART ON				
SEX	0.421	0.163	2.584	0.010
TRT	-0.536	0.192	-2.786	0.005
Between Level				
HEART ON				
HOURS	0.292	0.124	2.367	0.018

Intercepts				
HEART	0.851	0.141	6.020	0.000
Residual Variances				
HEART	1.393	0.235	5.919	0.000

Given the statistically significant and positive coefficient for hours (0.292), these results show that the more the number of hours for which a center is open, the more the warning signs that patients who attend will experience over a 6-month period. In other respects, these results do not differ substantially from those in Model 8.7, which, in turn, do not differ substantially from those in the earlier models, generally revealing similar relationships among the independent and dependent variables.

Negative Binomial Models for Multilevel Data

Recall that the signal quality of the Poisson distribution is the equality of the mean and variance. In some instances, however, the variance of a variable may be larger than the mean, leading to the problem of overdispersion, which we described in Chapter 7. As we learned in the previous chapter, the negative binomial distribution presents an alternative for use when the outcome variable is overdispersed. The negative binomial distribution takes an alternate form of the Poisson, with a difference in the variance parameter (see Chapter 7 for a discussion of this difference). In order to fit the negative binomial model with multilevel data by using Mplus, we can use the following program, where treatment and sex are the independent variables, as was the case in Model 8.7. The program is very similar to that used for Model 8.7, with the addition of (nb) in the count line, to indicate that we are fitting a negative binomial model.

```
TITLE: Model 8.10 Multilevel Negative Binomial regression
model
DATA:  FILE IS rehab.csv;
VARIABLE:    NAMES ARE heart trt sex hours rehab;
             count IS heart (nb);
         usevariables are heart trt rehab sex;
         cluster=rehab;
         within=trt sex;

ANALYSIS:
         type=twolevel;
MODEL:
         %within%
         heart ON trt sex;

         %between%
         heart;
```

```
OUTPUT:
        tech1 tech8 standardized;

savedata:
results are model810.out;
```

SUMMARY OF DATA

    Number of clusters   110

COUNT PROPORTION OF ZERO, MINIMUM AND MAXIMUM VALUES

     HEART   0.590   0   225

THE MODEL ESTIMATION TERMINATED NORMALLY

MODEL FIT INFORMATION

Number of Free Parameters   5

Loglikelihood

        H0 Value   -1963.153
        H0 Scaling Correction Factor   1.1420
          for MLR

Information Criteria

        Akaike (AIC)   3936.306
        Bayesian (BIC)   3960.845
        Sample-Size Adjusted BIC   3944.964
          (n* = (n + 2) / 24)

MODEL RESULTS

	Estimate	S.E.	Est./S.E.	Two-Tailed P-Value

Within Level

```
HEART        ON
    TRT              -0.480        0.182       -2.640        0.008
    SEX               0.380        0.196        1.942        0.052

Dispersion
    HEART             5.399        0.524       10.299        0.000

Between Level

Means
    HEART             1.266        0.138        9.149        0.000

Variances
    HEART             0.276        0.149        1.852        0.064
```

STANDARDIZED MODEL RESULTS

STDYX Standardization

	Estimate	S.E.	Est./S.E.	Two-Tailed P-Value

Within Level

```
HEART        ON
    TRT              -0.800        0.124       -6.437        0.000
    SEX               0.634        0.159        3.978        0.000

Dispersion
    HEART             0.000      999.000      999.000      999.000

Between Level

Means
    HEART             2.409        0.806        2.988        0.003

Variances
    HEART             1.000        0.000      999.000      999.000
```

These results are similar to those described above for Model 8.7, indicating the significant relationships between the frequency of cardiac warning signs and both treatment and sex, with the same signs as were present in the

earlier model. Mplus also provides us with the dispersion parameter estimate (5.399), which is much larger than 1, which is the value that we would
expect for the Poisson model. This relatively large value would support
the use of the negative binomial, as it indicates that the data are definitely
overdispersed.

Inclusion of Additional Level 2 Effects to the Multilevel Negative Binomial Regression Model

As with the Poisson regression model, we can include level 2 independent
variables, such as number of hours for which the centers are open.

```
TITLE: Model 8.11 Multilevel Negative Binomial regression
model with level 2 and level 2 covariates
DATA:  FILE IS rehab.csv;
VARIABLE:    NAMES ARE heart trt sex hours rehab;
             count IS heart (nb);
       usevariables are heart trt rehab sex hours;
       cluster=rehab;
       within=trt sex;
       between=hours;

ANALYSIS:
       type=twolevel;
MODEL:
       %within%
       heart ON trt sex;

       %between%
       heart;
       heart ON hours;
OUTPUT:
       tech1 tech8 standardized;

savedata:
results are model811.out;

SUMMARY OF DATA

    Number of clusters                          110


COUNT PROPORTION OF ZERO, MINIMUM AND MAXIMUM VALUES

    HEART       0.590           0         225
```

```
THE MODEL ESTIMATION TERMINATED NORMALLY


MODEL FIT INFORMATION

Number of Free Parameters                          6

Loglikelihood
        H0 Value                          -1959.948
        H0 Scaling Correction Factor        1.0999
          for MLR

Information Criteria

        Akaike (AIC)                       3931.896
        Bayesian (BIC)                     3961.342
        Sample-Size Adjusted BIC           3942.286
          (n* = (n + 2) / 24)



MODEL RESULTS
```

	Estimate	S.E.	Est./S.E.	Two-Tailed P-Value
Within Level				
HEART ON				
TRT	-0.468	0.169	-2.775	0.006
SEX	0.378	0.189	1.997	0.046
Dispersion				
HEART	5.416	0.532	10.179	0.000
Between Level				
HEART ON				
HOURS	0.256	0.093	2.750	0.006
Intercepts				
HEART	1.230	0.130	9.494	0.000
Residual Variances				
HEART	0.205	0.162	1.269	0.205

STANDARDIZED MODEL RESULTS

STDYX Standardization

	Estimate	S.E.	Est./S.E.	Two-Tailed P-Value
Within Level				
HEART ON				
TRT	-0.794	0.126	-6.330	0.000
SEX	0.642	0.158	4.063	0.000
Dispersion				
HEART	0.000	999.000	999.000	999.000
Between Level				
HEART ON				
HOURS	0.473	0.220	2.155	0.031
Intercepts				
HEART	2.393	0.865	2.767	0.006
Residual Variances				
HEART	0.776	0.208	3.733	0.000

As we have seen previously, the number of hours for which the centers are open is significantly positively related to the number of warning signs over the 6-month period of the study. In addition, individuals in the treatment condition had fewer heart warning signs and males had more heart warning signs than females. In addition, AIC, BIC, and aBIC are all lower in Model 8.11 than in Model 8.10, yielding evidence that including the number of hours for which the centers are open provides a better fit to the data.

Survival Analysis with Multilevel Data

In Chapter 7, we discussed models designed specifically for time until event data, whereby the dependent variable was the hazard rate conditioned on one or more independent variables, $h(t|x)$. Similar models can be fit in the context of multilevel data by using Mplus. As an example, let us consider a community wellness program in which 1000 patients, in a large urban area, who have been diagnosed with diabetes are invited to participate in a daily program of exercise. The program is hosted in one of 40 health clinics across the city. Data were collected over a 12-month period, with the outcome of interest being the number of days until a patient experienced an acute diabetic incident requiring hospitalization. The researchers are interested in whether there is a relationship between the number of days for which individual patients participated in the exercise program and the time until an acute incident was experienced, which is a level 1 variable. In addition, they want to know whether there is a relationship between time until an acute event and the number of staff members employed at each center, which is a level 2 variable.

Following is the Mplus code to fit a random intercept model for a 2-level survival analysis, based on the example above.

```
TITLE: Model 8.12  Cox regression with a random intercept
DATA:  FILE = multilevel_survival_data.dat;
VARIABLE:   NAMES = time days staff censored center;
       CLUSTER = center;
       WITHIN = days;
       BETWEEN = staff;
       SURVIVAL = time (ALL);
       TIMECENSORED = censored (0 = NOT 1 = RIGHT);
DEFINE:
   CENTER days (GRANDMEAN);
ANALYSIS:    TYPE = TWOLEVEL;
       BASEHAZARD = OFF;
MODEL: %WITHIN%
       time ON days;
       %BETWEEN%
       time ON staff;
       time;
```

The basic structure of the program combines elements of both the survival analysis code demonstrated in Chapter 7, and the multilevel modeling conventions discussed above. In this case, we have both level 1 and level 2 variables, with the number of days that the participants used the wellness facilities, and the number of staff members at each facility. The outcome is the time until an acute diabetes related event occurred. Those individuals who did not experience such an incident were considered to be right censored, and were coded as 1 in the censored variable. The output from this analysis appears below.

```
THE MODEL ESTIMATION TERMINATED NORMALLY
```

```
MODEL FIT INFORMATION
```

```
Number of Free Parameters                       3
```

```
Loglikelihood
        H0 Value                        77.930
        H0 Scaling Correction Factor     0.6966
          for MLR

Information Criteria

        Akaike (AIC)                    -149.860
        Bayesian (BIC)                  -135.137
        Sample-Size Adjusted BIC        -144.665
          (n* = (n + 2) / 24)
```

MODEL RESULTS

	Estimate	S.E.	Est./S.E.	Two-Tailed P-Value
Within Level				
TIME ON				
DAYS	0.502	0.038	13.092	0.000
Between Level				
TIME ON				
STAFF	−0.025	0.109	−0.230	0.818
Residual Variances				
TIME	0.414	0.090	4.627	0.000

From these results, we can see that individuals who spent more days in the wellness center also experienced a statistically significantly longer time until the advent of an acute diabetic event. On the other hand, there was no statistically significant relationship between the number of staff members at a center and the time until an event.

Summary

In this chapter, we learned that the generalized linear models featured in Chapter 7, which accommodate categorical dependent variables as well as time until event scenarios, can be easily extended to the multilevel context. Indeed, the basic concepts that we learned in Chapter 2 regarding sources of variation and various types of models can be easily extended for categorical outcomes. In addition, Mplus provides for easy fitting of a wide variety of such models. Therefore, in many ways, Chapter 8 represents a review of material that by now should be familiar to us, even while applied in a different scenario than we have seen up to now. Perhaps, the most importance point to take away from this chapter is the notion that modeling multilevel data in the context of generalized linear models is not radically different from the normally distributed continuous dependent variable case, so that the same types of interpretations can be made and the same types of data structure can be accommodated.

9

Brief Review of Latent Variable Modeling in Mplus

In the previous chapters, we have examined multilevel models involving variables of various types, including normally distributed continuous, ordinal, dichotomous, and time-until event. In all of these situations, the variables in question were assumed to be directly observed or measured. In many areas of research, including social sciences such as psychology, education, and business, the variables that researchers are most interested in are frequently not directly observable. These unobserved, or latent, variables include constructs such as depression, reading aptitude, intelligence, and personality traits such as introversion, to name only a few. Although these latent variables cannot be directly observed, researchers are able to use proxy measures for them, often in the form of questionnaires. The items, or subscales, on such instruments provide information about the construct of interest (e.g., intelligence). A variety of statistical techniques have been developed for the purpose of estimating such latent variables and, subsequently, relating them to one another, much in the same way that observed variables are related to one another by using regression. In addition, latent variable techniques exist for the purpose of identifying subgroups of individuals within the population and for modeling change over time in continuous and categorical latent variables. In this chapter, we briefly introduce several of these models in the single-level context and demonstrate how they can be fit by using Mplus. In the next chapter, we will extend these models to the multilevel data case.

Factor Analysis

Factor analysis (FA) is very widely used across many research disciplines. FA models link a set of observed variables (e.g., items on a scale and subscales from an assessment battery) to one or more latent traits that the observed variables are believed to measure. A primary assumption underlying these models is that, although the latent variable cannot be directly observed, it does have a direct influence on the observed variables. As we will discuss below, the relationships between the observed and unobserved variables can be modeled, and estimates of the strength and nature of the relationships between the two can be obtained in the form of what are known as factor loadings. FA can be conceived as either exploratory factor analysis (EFA) or confirmatory factor analysis (CFA) in nature. When using EFA, the researcher does not assume any specific structure on the model linking the latent variable(s) with the observed indicators but rather allows the nature of this structure to be derived from the data itself. This is not to say that the researcher using EFA must have no preconceptions regarding the latent structure underlying the data. In fact, the use of such models will typically go more smoothly if she does have some a priori ideas about the number of factors that are likely to be present and about the indicators that should go together on the same factor. However, no direct links are specified by the researcher between indicators and factors. In contrast to EFA, which derives the latent structure from the data, CFA models are built upon constraints imposed by the researcher based on theoretical links between the indicators and the latent variables. These constraints directly express which variables the researcher believes will be associated with which factors, and the proposed model is then assessed to ascertain how well it fits the observed data. In many instances, more than one CFA model are fit to the data, and the fits of these models are then compared with one another based on a variety of statistical and theoretical criteria. All these models should have some theoretical basis, and the one that provides the best statistical fit and has a solid theoretical basis is then selected as optimal.

With respect to when one should use EFA or CFA, the researcher must consider how much theory and prior research exist with respect to the latent structure underlying the observed data. When there exists a strong, well-established theoretical basis underlying one or more proposed latent variable models, coupled with prior exploratory work providing further support for these models, CFA would be the appropriate technique to use. On the other hand, when either solid theory based in the literature does not exist or there is little existing empirical support for the latent structure underlying the observed data, EFA may be most appropriate. EFA would be optimal in this latter case, because it places a relatively small burden on the researcher regarding the expected nature of the latent constructs and their relationships with the observed indicators. In the following pages, we will explore the modeling of indicators in the EFA framework by using Mplus. We will also learn how to interpret the results from such an analysis.

Exploratory Factor Analysis

The basic FA model can be expressed as

$$y = \Lambda\eta + \varepsilon \tag{9.1}$$

where:
y is the vector of observed indicator variables
η is the latent variable(s)

Λ are the factor loadings linking observed indicators with factors
ε are the unique variances for the indicators

The ε are assumed to be independent of one another and of the factors in the model. The Λ, which typically will range between −1 and 1, reflect the relationships between factors and indicators, with larger values being indicative of a closer association between the two. Parameters from Model Equation 9.1 can be used to predict the covariance matrix of the indicator variables, as expressed in Equation 9.2.

$$\Sigma = \Lambda\Psi\Lambda' + \Theta \tag{9.2}$$

where:
  Σ is the model-predicted correlation matrix of the indicators
  Ψ is the correlation matrix for the factors
  Θ is the diagonal matrix of unique error variances

Factor Extraction

EFA involves two primary steps: (1) the initial estimation of the parameters in Equations 9.1 and 9.2, known as factor extraction, and (2) the transformation of Λ to aid in the interpretability of the solution, or factor rotation. A number of factor extraction methods are available, including, perhaps, the two most popular approaches, maximum likelihood (ML) and principal axis factoring (PAF). The goal of these techniques is to find estimates for the parameters in Models Equations 9.1 and 9.2 that will results in Σ being as close to the observed indicator variable covariance matrix, **S**, as possible. The ML approach is fit based on the assumption of multivariate normality for the observed indicators, whereas PAF does not make such an assumption.

Factor Rotation

The second major aspect in conducting an EFA is the rotation of the factor loadings. When multiple factors are extracted, the model in Equation 9.1 is indeterminate in nature, meaning that there are an infinite number of combinations for Λ that will yield the same estimate of Σ. Given this fact, interpretability of the solution serves as an important determinant in the final form of the model, where interpretability refers to the extent to which the factor loadings clearly link each observed variable to (ideally) a single latent variable. Another way in which the interpretability of the solution can be considered is the extent to which it achieves simple structure, which Thurstone (1947) defined as occurring when the following conditions were met: (1) Each factor has associated with it a subset of the indicator variables with which it is highly associated (i.e., large loadings). (2) Each indicator is highly associated with only one factor and has loadings near 0 on the other factors.

Simple structure makes the results of the model easier to interpret. However, typically, the initial Λ extracted from the data does not meet the conditions necessary for simple structure to hold. Factor rotation is used to transform the loadings in order to achieve a solution that is as close to simple structure as possible. It does not change the actual fit of the model (i.e., the proximity of Σ and S) but simply transforms the values of Λ to better approximate simple structure.

Factor rotation methods are either orthogonal or oblique, where orthogonal rotations constrain the correlations among factors to be 0 and oblique rotations allow the factors to be correlated. Within both broad types of rotations, there are a number of options available to the researcher. No one of these rotations is optimal in every research scenario, though some are more popular than others, such as Varimax among the orthogonal rotations and Promax among the oblique rotations. It should be noted, however, that there are many others to choose from, and some, such as Geomin, have been shown to be particularly effective in a variety of situations (Finch, 2011). The decision regarding the type of rotation to use should be based on both theory regarding the nature of the constructs under study and empirical evidence about the correlation structure of the factors. It may be advisable for the researcher to initially use an oblique rotation such as Promax, particularly if he believes that the factors should be correlated with one another. If the estimated correlations among the latent variables are small, then the EFA could be fit to the data again and an orthogonal rotation can be used. Otherwise, the results from the initial oblique rotation could serve as the final estimates of Λ.

Determining the Number of Factors to Retain

Researchers using EFA will typically fit several models to the data, each differing in terms of the number of factors that are retained. The primary question of interest in such cases is, what is the optimal model in terms of the number of factors? There are several statistical tools available for determining the number of factors to retain, and these are typically used in conjunction with one another, along with theoretical considerations regarding the plausibility of the identified factors, to ascertain the optimal solution. Some of the statistical approaches for identifying the best factor solution have been found to work better than others, but none of them has been shown to be universally best in all cases. Among the earliest statistical methods for determining the number of factors to retain involves an examination of the eigenvalues associated with each of the factors (Kaiser, 1960). Eigenvalues reflect the amount of variance associated with each of the factors. Furthermore, when the indicators are standardized, they have a variance of 1. Thus, factors that have eigenvalues greater than 1 account for more variance than the individual indicators and are therefore retained as being important. Although it has a certain intuitive appeal, the eigenvalue-greater-than-1 approach often results in the retention of too many factors (Pett, Lackey, & Sullivan, 2003).

The scree plot is a second statistical approach that uses the eigenvalues to determine the number of factors to retain. The scree plot is a graphical method in which the eigenvalues for the factors are placed on the *y*-axis and the factor numbers are placed on the *x*-axis. The researcher simply examines the plot, looking for the place in the data where the difference in the eigenvalues between adjacent factors declines. A third descriptive tool for deciding on the number of factors to retain involves an examination of the residual correlation matrix for the observed indicators. As discussed above, in ML estimation of the model, parameters for Equations 9.1 and 9.2 were arrived at by minimizing the difference between the model-predicted and actual covariance matrices for the indicator variables. Thus, a model that fits the data well is one that yields Σ close to **S**. Given that the standardized version of Σ is the correlation matrix, the researcher can examine residual correlations (the difference between the observed and model-predicted correlations) for the indicators in order to ascertain the goodness of fit for the model. By convention, a good EFA solution is one with few residual correlations greater than 0.05 (Gorsuch, 1983).

In addition to the descriptive methods for determining the number of factors to retain, there are also inferential approaches for this purpose. These techniques involve statistical tests of model fit for a specific number of factors, for example, whether a model with $m + 1$ factors fits better than a model with m factors and whether the eigenvalue for the mth factor using the observed data is larger than the eigenvalue for the mth factor based on data for which it is known that no factor structure is present. When ML is used for factor extraction, a χ^2 test can be calculated to assess the fit of a given model to the data by comparing the relative proximity of Σ to **S**. This statistic is calculated as

$$\chi^2 = \Big[\ln S - \ln \Sigma + \text{trace}\Big[S\big(\Sigma^{-1}\big)\Big] - p \Big](N-1) \tag{9.3}$$

The terms in Equation 9.3 are as defined before, with the addition that p is the number of indicator variables and N is the total sample size. The null hypothesis being tested is $\Sigma = S$. Such perfect fit is rarely achieved, even for sample models that are reasonably close to the population generating model, leading to frequent rejection of the null hypothesis, even when the model may represent a good approximation of the observed data. For this reason, the χ^2 test is not seen as especially helpful in determining the number of factors to retain in EFA, particularly for larger sample sizes (Kim & Muller, 1978).

An alternative use for the χ^2 statistic when identifying optimal EFA models is the calculation of a difference of values for two competing models. In other words, if the researcher is interested in comparing the relative fit of m and $m + 1$ factor models, she can calculate the difference in χ^2 statistics for the two models and interpret the resulting value as a χ^2 itself, testing the null hypothesis that the fit of the two models to the data is equivalent. This difference test is appropriate as long as the two models are nested within one another (i.e., one model is a more constrained version of another). In that case, the difference test is calculated as

$$\chi^2_\Delta = \chi^2_m - \chi^2_{m+1} \tag{9.4}$$

where:

χ^2_m is the χ^2 statistic for model with m factors

χ^2_{m+1} is the χ^2 statistic for model with $m + 1$ factors

The degrees of freedom for this difference test are equal to the difference in degrees of freedom for the two models. When χ^2_Δ is statistically significant at the desired level of α (e.g., 0.05), the researcher would reject the null hypothesis of equivalent fit for the two models and select the simpler of the two models as optimal.

A third inferential approach that has been recommended for use in ascertaining the number of factors to retain is parallel analysis. This methodology was first described by Horn (1965) and is conducted in the following manner:

1. Fit an EFA to the original dataset and retain the eigenvalues for each factor.

2. Generate observed data with marginal characteristics identical to the observed data (i.e., same means and standard deviations) but with uncorrelated indicators, either by randomly sorting the values of the observed indicators or by simulating such data.

3. Fit an EFA to the generated data and retain the eigenvalues for each factor.

4. Repeat steps 2 and 3 many (e.g., 1000) times in order to develop distributions for each eigenvalue under the case where indicators are not related to one another.

5. Compare the observed eigenvalue for the first factor with the 95th percentile from the distribution of first-factor eigenvalues of the generated data. If the observed value is greater than or equal to the 95th percentile value, conclude that at minimum, one factor should be retained, and continue to step 6. If the observed value is less than the 95th percentile, then stop and conclude that no common factors exist.

6. Compare the observed eigenvalue for each successive factor with the 95th percentile for the corresponding eigenvalue distribution of the generated data. If the observed value is greater than or equal to the 95th percentile, retain that factor (e.g., the second factor and the third factor) and move to the next factor in line. This process stops when the observed eigenvalue is less than the 95th percentile of the generated data.

Parallel analysis has been found to be one of the most accurate methods for identifying the number of factors to retain in an EFA (Fabrigar & Wegener, 2011). Once again, however, it must be emphasized that no single approach for determining the number of factors to retain is perfect, and thus, we suggest that the researcher use several of these methods in conjunction with one another.

Exploratory Factor Analysis by Using Mplus

In order to demonstrate the fitting of EFA by using Mplus, let us consider the following example. A researcher has collected data on a 30-item scale measuring perfectionism, from 784 individuals. It is hypothesized that the 30 items can be grouped into four factors that measure different aspects of perfectionism. Theoretically, items 1–6 are associated with one factor, items 7–17 with a second factor, items 18–24 with a third factor, and items 25–30 with a fourth factor. The data are contained in the file model9_1.dat, and the item responses are represented as standardized values. The following Mplus program fits EFA models from 1 to 5 factors and requests that parallel analysis be conducted on 1000 random datasets.

```
TITLE: Model 9.1  EFA

DATA: FILE IS model9_1.dat;

VARIABLE: NAMES ARE score1-score30;
missing are .;

ANALYSIS: TYPE=EFA 1 5;
estimator=ml;
parallel=1000;

plot:   type=plot2;

output: residual;
```

Aspects of this program are similar to those that were used in earlier chapters and thus will not be described here. However, there are sections that are unique to the fitting of EFA. For example, under ANALYSIS:, we indicate that the model to be fit is EFA and that we would like Mplus to provide results for 1, 2, 3, 4, and 5 factors, as indicated by TYPE=EFA 1 5. The default extraction technique is maximum likelihood, so strictly speaking, we would not need to explicitly indicate this by using the estimator=ml subcommand. In addition, we have requested that parallel analysis be done with 1000 random datasets, as mentioned above (parallel=1000). In order to obtain the scree plot, as well as a graphical representation of the parallel analysis results, we must specify type=plot2 and that residual correlations are obtained with output: residual;.

When interpreting the results of the EFA, we must first determine the number of factors to retain, after which we can examine the factor loadings for this optimal solution. Theory would suggest that we should retain four factors. However, we must also consider what the empirical evidence indicates before making our final determination. Below are the results for both the χ^2 and χ^2_Δ tests.

SUMMARY OF MODEL FIT INFORMATION

Model	Number of Parameters	Chi-Square	Degrees of Freedom	P-Value
1-factor	90	5919.494	405	0.0000
2-factor	119	3851.177	376	0.0000
3-factor	147	2060.168	348	0.0000
4-factor	174	375.592	321	0.0193
5-factor	200	329.682	295	0.0804

Models Compared	Chi-Square	Degrees of Freedom	P-Value
1-factor against 2-factor	2068.317	29	0.0000
2-factor against 3-factor	1791.008	28	0.0000
3-factor against 4-factor	1684.577	27	0.0000
4-factor against 5-factor	45.909	26	0.0093

These results suggest that the 5-factor model does provide reasonable fit to the data, given the nonsignificant χ^2 value ($p = 0.0804$). In addition, the 5-factor model fits the data significantly better than the 4-factor solution ($p = 0.0093$). The scree plot and parallel analysis results appear in Figure 9.1.

Based on the parallel analysis, it would appear that we should retain four factors, as the 95th percentile of the 5th eigenvalue for the distribution of random data EFA results is greater than the observed 5th eigenvalue, but such is not the case for the 4th eigenvalue. In addition, the scree plot shows a clear flattening of the line linking the eigenvalues at the 5th factor. The eigenvalues for the first 10 factors appear below.

EIGENVALUES FOR SAMPLE CORRELATION MATRIX

	1	2	3	4	5
1	5.981	4.046	3.528	3.298	0.685

EIGENVALUES FOR SAMPLE CORRELATION MATRIX

	6	7	8	9	10
1	0.675	0.661	0.629	0.615	0.613

The first four eigenvalues are larger than 1, but the fifth is not, which would also suggest that four factors should be retained.

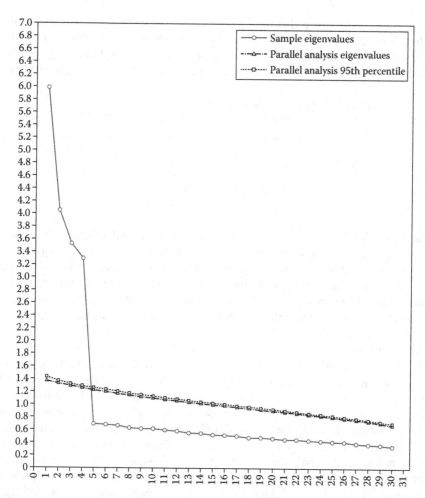

FIGURE 9.1
Scree plot and parallel analysis results for the perfectionism data.

Finally, we can examine the residual correlations, which can be a tedious process, given the large number of residual correlations that exist for many research scenarios. As an example, below are the residual correlations for the first five indicators for the 1-factor solution.

```
          Residuals for Correlations
               SCORE1        SCORE2        SCORE3        SCORE4        SCORE5
              _____      _____      _____      _____      _____
SCORE1         0.000
SCORE2         0.482         0.000
SCORE3         0.498         0.510         0.000
SCORE4         0.511         0.542         0.507         0.000
SCORE5         0.465         0.519         0.486         0.502         0.000
```

All of these residuals in this brief example are larger than the 0.05 threshold value. Indeed, a similar pattern is present for all of the variable pairs, indicating that the 1-factor model offers poor fit to the data. Following are the residual correlations for the same first-five indicators for the 4-factor solution.

```
Residuals for Correlations
         SCORE1       SCORE2       SCORE3       SCORE4       SCORE5
         _____       _____       _____       _____       _____

SCORE1    0.000
SCORE2   -0.013        0.000
SCORE3    0.018        0.004        0.000
SCORE4    0.006        0.010       -0.010        0.000
SCORE5   -0.013        0.015       -0.005       -0.012        0.000
```

The difference in residual correlation values for the 1- and 4-factor solutions is clear, with none of those obtained using the 4-factor model being larger than the 0.05 cutoff. An examination of the full set of residuals reveals essentially the same result, indicating that the 4-factor solution provides good fit to the data. It should be noted here that the residual correlations from the 3-factor model were also generally below the cutoff of 0.05, with the exception of items 25 through 30.

Given these results, and theory, the question that must now be answered is, what is the optimal number of factors to retain? Recall that theory suggests the presence of 4-factor solution. However, the two χ^2 test results suggested that the 5-factor solution might provide a better fit to the data. On the other hand, both parallel analysis and the scree plot indicated that four factors were sufficient. In addition, the residual correlations for the 4-factor model were essentially all below the 0.05 threshold, thereby indicating the good fit of the model to the data. Considering theory together with the full set of fit statistics, we conclude that the 4-factor model is optimal. To be completely sure that this is the correct decision, we can also examine the factor loadings for the two different solutions. Below are the loadings for the 4-factor model, followed immediately by those from the 5-factor solution.

```
GEOMIN ROTATED LOADINGS (* significant at 5% level)
                 1            2            3            4
               _____       _____       _____       _____

SCORE1         0.681*       0.026       -0.035       -0.021
SCORE2         0.720*       0.004       -0.003       -0.022
SCORE3         0.708*      -0.078*       0.036        0.035
SCORE4         0.735*       0.013       -0.008       -0.018
SCORE5         0.698*      -0.036        0.014       -0.011
SCORE6         0.712*       0.052*      -0.022        0.029
SCORE7         0.034        0.725*       0.010        0.037
SCORE8         0.002        0.710*      -0.001       -0.017
SCORE9        -0.046        0.673*      -0.025       -0.009
SCORE10       -0.003        0.688*       0.001       -0.026
```

SCORE11	0.006	0.696*	-0.029	0.017
SCORE12	0.020	0.688*	0.029	0.025
SCORE13	0.027	0.702*	0.024	-0.044
SCORE14	-0.016	0.718*	0.001	0.029
SCORE15	0.031	0.728*	0.023	-0.048
SCORE16	-0.042	0.686*	-0.012	0.045
SCORE17	-0.028	0.699*	-0.010	-0.019
SCORE18	0.029	0.056*	0.662*	0.008
SCORE19	-0.030	-0.017	0.719*	0.010
SCORE20	0.043	0.009	0.719*	0.009
SCORE21	-0.037	-0.029	0.683*	-0.002
SCORE22	0.022	-0.012	0.694*	-0.070*
SCORE23	0.003	0.015	0.714*	0.025
SCORE24	-0.055*	-0.001	0.676*	0.005
SCORE25	-0.013	-0.032	-0.029	0.699*
SCORE26	-0.026	-0.029	-0.019	0.704*
SCORE27	-0.005	0.032	0.010	0.703*
SCORE28	0.020	0.033	0.007	0.697*
SCORE29	0.079*	-0.012	0.015	0.699*
SCORE30	-0.036	0.005	0.018	0.699*

GEOMIN ROTATED LOADINGS (* significant at 5% level)

	1	2	3	4	5
SCORE1	0.681*	0.028	-0.034	-0.021	0.019
SCORE2	0.749*	-0.010	0.004	-0.024	0.171*
SCORE3	0.700*	-0.069*	0.032	0.039	-0.058
SCORE4	0.740*	0.012	-0.007	-0.018	0.048
SCORE5	0.688*	-0.022	0.008	-0.006	-0.113
SCORE6	0.707*	0.059*	-0.024	0.032	-0.035
SCORE7	0.039	0.721*	0.012	0.035	0.044
SCORE8	-0.009	0.719*	-0.005	-0.017	-0.082
SCORE9	-0.039	0.668*	-0.023	-0.011	0.057
SCORE10	-0.001	0.685*	0.002	-0.027	0.027
SCORE11	0.012	0.691*	-0.027	0.015	0.051
SCORE12	0.019	0.688*	0.028	0.023	-0.011
SCORE13	0.020	0.708*	0.022	-0.044	-0.049
SCORE14	-0.024	0.724*	-0.002	0.030	-0.058
SCORE15	0.032	0.725*	0.023	-0.049	0.018
SCORE16	-0.015	0.669*	-0.002	0.039	0.243*
SCORE17	-0.037	0.706*	-0.014	-0.019	-0.075
SCORE18	0.039	0.046	0.668*	0.005	0.090
SCORE19	-0.032	-0.017	0.718*	0.010	-0.012
SCORE20	0.045	0.005	0.720*	0.009	0.023
SCORE21	-0.038	-0.029	0.682*	-0.003	-0.013
SCORE22	0.020	-0.012	0.693*	-0.070*	-0.023
SCORE23	-0.005	0.020	0.712*	0.027	-0.065
SCORE24	-0.053	-0.004	0.677*	0.003	0.021
SCORE25	-0.009	-0.039	-0.026	0.698*	0.063
SCORE26	-0.036	-0.026	-0.021	0.702*	-0.062

SCORE27	-0.009	0.029	0.010	0.701*	0.006
SCORE28	0.017	0.032	0.007	0.694*	-0.001
SCORE29	0.053	0.007	0.006	0.716*	-0.214
SCORE30	-0.026	-0.008	0.023	0.700*	0.112

First, we should note that the rotation used was Geomin, which is the default in Mplus. Second, Mplus automatically conducts a hypothesis test for each of the factor loadings, against the null hypothesis that the loading is 0. Those that are significantly different from 0 are flagged with an *. The test statistic is simply the loading divided by its standard error. Mplus displays both the standard errors and the test statistics in tables immediately below the rotated loadings. In addition to the significance test, it is also common for researchers to use some cutoff value (e.g., 0.3) to identify loadings that are considered nontrivial (Tabachnick & Fidell, 2013). An examination of the results for the 4-factor model reveals that the items do group with the loadings as theory would suggest they should, using either the hypothesis test results or the cutoff value of 0.3. In addition, there are no cross-loadings that exceed 0.3, meaning that none of the items were associated with multiple factors.

In contrast to the 4-factor model, factor loading results were not as supportive of the 5-factor solution. In particular, although the items loaded together in the theoretically expected fashion, the 5th factor did not have any variables for which the loadings exceeded the 0.3 value. Thus, the 5th factor appears to be unnecessary and would lead the researcher to conclude that the 4-factor model is optimal. Therefore, the results of this EFA do indeed support the theoretically proposed latent structure, both in terms of the number of latent variables and the way in which the observed indicators associate themselves with those indicators. As mentioned earlier, the default rotation used by Mplus is the oblique Geomin. However, the researcher can change this rotation by inserting the following command under `analysis`:

```
rotation=promax;
```

Mplus has a wide array of rotations available, and the interested reader is encouraged to investigate these.

Confirmatory Factor Analysis

In the previous section, we described the EFA model, which links observed indicator variables with one or more latent variables, without the researcher imposing any constraints on how these relationships might be formed. As also discussed briefly earlier in this chapter, there exists a second type of factor analysis, CFA, which does place constraints on the factor model that are not present in EFA. These constraints typically (though not always) involve setting factor loadings to 0 for combinations of indicator variables and factors, so that each indicator is linked only to a single factor. The factor model presented in Equation 9.1 is representative of both EFA and CFA, although researchers often represent CFA slightly differently, as in Equation 9.5.

$$y = \tau + \Lambda\eta + \varepsilon \tag{9.5}$$

In CFA models, τ (the intercept) is not always assumed to be 0, unlike with EFA. The intercept comes into play when researchers are interested in comparing the means of latent variables among multiple groups in the data.

Although the same basic model underlies both EFA and CFA, the two techniques do differ from one another in some important ways. EFA is descriptive in nature, and as such, no presuppositions are made about the underlying latent structure. On the other hand, the researcher using CFA will fit one or more specific factor models that are differentiated by the constraints that are employed, that is, which indicators are assumed to be associated with which factors. For this reason, the considerations that the researcher brings to the analysis are different for CFA than for EFA. Of particular interest, for example, is the question of model fit. In other words, how well does the proposed model represent the observed data? While this issue is certainly considered in the context of EFA, the larger question in that case involves the number of factors to retain and whether simple (or approximately simple) structure can be achieved. In CFA, the model is introduced a priori, and its ability to accurately reproduce the observed covariance matrix is of primary interest. Below, we provide a brief introduction to CFA, focusing, in particular, on fitting these models by using Mplus.

Fitting CFA Models and Assessing Goodness of Fit

There are a number of algorithms available for fitting CFA models. Certainly, among the most popular of these, and frequently, the default in software packages, is ML, which has been discussed above in the context of EFA. ML assumes that the observed indicators are multivariate normal and can yield poor estimates of the standard errors if this is not the case (Yuan, Bentler, & Zhang, 2005). One widely used alternative is weighted least squares (WLS), which takes a least squares approach to estimating model parameters and applies a weight such that variances and covariances with lower sampling variability will be given more weight in the estimation of CFA model parameters. While this method has been shown to work quite well for large samples, it has somewhat limited utility for small samples, owing to the computational complexity of the algorithm (Finney & Distefano, 2013). An alternative to WLS involves using only the diagonal elements of the weight matrix from WLS and is thus known as diagonally weighted least squares (DWLS) estimation. It has been shown to work well under a variety of conditions, particularly with samples as small as 100 (Wirth & Edwards, 2007; Flora & Curran, 2004; Muthén, du Toit, & Sipsic, 1997).

Once parameters have been estimated, the researcher must determine the extent to which the proposed model fits the observed data. There are a large number of methods available for this purpose, including the χ^2 goodness of fit test, which was described earlier in the context of EFA. It is used in exactly the same manner with CFA as was done in the case for EFA, in order to test the same null hypothesis: $\Sigma = S$. However, as noted earlier, this is a very restrictive null hypothesis, and the test statistic is likely to be significant even for very small deviations between Σ and S, particularly when the sample size is large (Bollen, 1989). There also exist several corrections to the basic χ^2 test that can be used when the data are not normally distributed (Bollen and Stine, 1992; Satorra & Bentler, 1994; Yuan and Bentler, 1997). These have been shown to work better than the standard test statistic when the normality assumption is violated and each of these adjustments is available in Mplus.

The χ^2 statistic tests the null hypothesis of exact model fit. There are, however, other statistics for the purpose of determining the statistical plausibility of a CFA model that are more relative in nature. Each of these provides information regarding the degree to which the model fits the data, not whether the model fits the data with absolute perfection. Researchers can use several of these indices and consider the collective results in order to come to a decision regarding the fit of the model. There are several such indices, and we will describe those that research has proven to work well in many situations and that are provided in the Mplus output. One such index is the Root Mean Square Error of Approximation (RMSEA), which is calculated as

$$\text{RMSEA} = \sqrt{\frac{\chi_T^2 - df_T}{df_T(n-1)}} \tag{9.6}$$

where:

χ_T^2 is the ML-based χ^2 test for the target model (i.e., the model of interest)

df_T is the degrees of freedom for the target model (number of observed covariances and variances minus number of parameters to be estimated)

n is the sample size

Values of RMSEA $\leq$ 0.05 indicate good model fit, whereas values between 0.05 and 0.08 can be interpreted as suggesting adequate model fit (Kline, 2016). Values of RMSEA greater than 0.08 indicate poor fit for the model.

The comparative fit index (CFI) is a second widely used and effective measure of model fit.

$$\text{CFI} = 1 - \frac{\text{Max}\left(\chi_T^2 - df_T, 0\right)}{\text{Max}\left(\chi_0^2 - df_0, 0\right)} \tag{9.7}$$

where:

χ_0^2 is the ML-based χ^2 test for the null model in which no relationships between the latent and observed variables are hypothesized to exist

df_0 is the degrees of freedom for the null model

A closely related fit statistic to the CFI is the Tucker-Lewis Index (TLI).

$$TLI = \frac{(\chi_0^2/df_0) - (\chi_T^2/df_T)}{(\chi_0^2/df_0) - 1} \tag{9.8}$$

A variety of studies have been conducted to identify guidelines for using CFI and TLI to identify good-fitting models. Hu and Bentler (1999) conducted a thorough simulation study focusing on this issue and suggested that in conditions similar to those that they simulated, CFI and TLI values of 0.95 or higher suggest good fit. It is important to keep in mind, however, that such cutoff values are only guidelines and that other values have been suggested (e.g., 0.9 and higher, indicating good fit). Another commonly used index of model fit is the Standardized Root Mean Square Residual (SRMR).

$$SRMR = \frac{\sum \left(r_{observed\ j,k} - r_{predicted\ j,k} \right)^2}{\left(p\ (p+1)/2 \right)} \tag{9.9}$$

where:

$r_{observed\ j,k}$ is the observed correlation between indicator variables j and k

$r_{predicted\ j,k}$ is the CFA-model-predicted correlation between indicator variables j and k

p is the number of observed indicator variables

Hu and Bentler (1999) suggested values of SRMR ≤ 0.08 suggest good model fit to the data.

In addition to assessing the fit of a single model, it is also often of interest to compare the fit of two models. This can be done by using χ_Δ^2, as was discussed in the context of EFA, in Equation 9.4. In the context of CFA, two nested models can be compared with one another, such that one of the models is a constrained version of the other. In that case, the test statistic would be expressed as

$$\chi_\Delta^2 = \chi_C^2 - \chi_U^2 \tag{9.10}$$

where:

χ_C^2 is the χ^2 statistic from constrained model

χ_U^2 is the χ^2 statistic from unconstrained model

The null hypothesis being tested is that the constrained model provides equivalent fit to the data as the unconstrained model. Degrees of freedom for the test are calculated as the difference in degrees of freedom for the two models. Model fit can also be compared by using the information indices that have been discussed in prior chapters, including Akaike's information criterion (AIC), Bayesian information criterion (BIC), sample size adjusted BIC (aBIC). Just as in these other contexts, models with smaller values of each of these statistics are said to fit the data better.

In addition to indices of model fit, the researcher should also consider the parameter estimates when evaluating whether a model fits the data. For example, regardless of the model fit information, if a number of the factor loading estimates are not significantly related to the latent variables in the manner that we would expect, then the model should not be judged to be appropriate. Focusing only on the fit indices could lead the researcher to ignore other important information suggestive of problems with the model. Thus, the researcher should examine the factor loadings, variances, and covariances, for example, to be sure that they appear to be appropriate in terms of magnitude and range. In addition, the model residuals should also be examined, much as was the case with EFA. Taken together with the model fit information, such an examination can provide the researcher with a full picture of the model fit.

Confirmatory Factor Analysis by Using Mplus

Fitting CFA models by using Mplus is very straightforward. Let us reconsider the perfectionism scale that was described above in the context of EFA. If there is strong theory coupled with prior empirical evidence (perhaps, in the form of EFA results published in the literature) to support the 4-factor model, the researcher may elect to fit a CFA to the data. Recall that theory suggests that items 1 through 6 are associated with factor 1, items 7 through 17 with factor 2, items 18 through 24 with factor 3, and items 25 through 30 with factor 4. The following Mplus program can be used to fit the CFA model with ML estimation, which is the default.

```
TITLE: Model 9.2  CFA of multidimensional perfectionism scale

DATA: FILE IS model9_1.dat;

VARIABLE: NAMES ARE score1-score30;

missing are .;

MODEL:        f1 BY score1-score6;
              f2 BY score7-score17;
              f3 by score18-score24;
              f4 by score25-score30;

output: standardized;
```

Much of this program mirrors earlier ones that we have examined, and thus, our focus here will be on the aspects that are new. In particular, note that unlike with EFA, in CFA, we explicitly link each indicator with each factor, as in f1 BY score1-score6;. In addition, we have requested standardized output, which means that we will obtain both the unstandardized and the standardized factor loadings. By default, Mplus uses the first variable in the list associated with each factor (e.g., score1 and score2) as the referent indicator to provide the latent variable with scale (Brown, 2015). It is also possible for the researcher to set the factor variance to 1 in order to provide scale and thereby override the default in Mplus of using the first indicator for this purpose, if that is so desired. Selected portions of the resulting output from the program appears below.

```
THE MODEL ESTIMATION TERMINATED NORMALLY


MODEL FIT INFORMATION

Number of Free Parameters                        96

Loglikelihood

           H0 Value                        -28779.153
           H1 Value                        -28542.588

Information Criteria

           Akaike (AIC)                     57750.306
           Bayesian (BIC)                   58198.089
           Sample-Size Adjusted BIC         57893.241
            (n* = (n + 2) / 24)

Chi-Square Test of Model Fit

           Value                             473.130
           Degrees of Freedom                    399
           P-Value                            0.0062

RMSEA (Root Mean Square Error Of Approximation)

           Estimate                            0.015
           90 Percent C.I.                     0.009  0.021
           Probability RMSEA <= .05            1.000

CFI/TLI

           CFI                                 0.992
           TLI                                 0.991
```

```
Chi-Square Test of Model Fit for the Baseline Model

         Value                          9871.676
         Degrees of Freedom                  435
         P-Value                          0.0000

SRMR (Standardized Root Mean Square Residual)

         Value                             0.028
```

Using the guidelines described above, we can conclude that the proposed model yields good fit to the data. The CFI and TLI are greater than 0.99, the RMSEA is 0.015, and the upper bound on the confidence interval, 0.021, is well below the 0.05 threshold, indicating good model fit. Likewise, the SRMR is well below the 0.08 cutoff. The χ^2 test is statistically significant, though, as noted above, it is very sensitive for larger samples, which we do have in this case ($N = 784$).

The unstandardized and standardized model parameter estimates appear below. Of most interest here is the extent to which the factor loadings are significantly related to the factors to which they have been assigned. It is expected that they all should be statistically significant if the model specification is correct. In addition to the loadings, the model covariances/correlations, model intercepts, and factor and error variances are also presented.

MODEL RESULTS

	Estimate	S.E.	Est./S.E.	Two-Tailed P-Value
F1 BY				
SCORE1	1.000	0.000	999.000	999.000
SCORE2	0.999	0.057	17.559	0.000
SCORE3	1.018	0.059	17.152	0.000
SCORE4	1.100	0.061	17.952	0.000
SCORE5	1.036	0.061	16.943	0.000
SCORE6	1.031	0.059	17.392	0.000
F2 BY				
SCORE7	1.000	0.000	999.000	999.000
SCORE8	0.990	0.051	19.305	0.000
SCORE9	0.865	0.048	18.176	0.000
SCORE10	0.972	0.052	18.585	0.000
SCORE11	0.945	0.050	18.795	0.000
SCORE12	0.916	0.049	18.685	0.000
SCORE13	0.988	0.052	19.137	0.000
SCORE14	0.975	0.050	19.518	0.000
SCORE15	1.018	0.051	19.895	0.000

SCORE16	0.930	0.050	18.544	0.000
SCORE17	0.936	0.050	18.883	0.000

F3   BY

SCORE18	1.000	0.000	999.000	999.000
SCORE19	1.135	0.066	17.141	0.000
SCORE20	1.097	0.064	17.084	0.000
SCORE21	1.063	0.065	16.340	0.000
SCORE22	1.069	0.065	16.554	0.000
SCORE23	1.098	0.064	17.077	0.000
SCORE24	1.083	0.066	16.379	0.000

F4   BY

SCORE25	1.000	0.000	999.000	999.000
SCORE26	1.011	0.058	17.443	0.000
SCORE27	1.025	0.060	17.208	0.000
SCORE28	1.009	0.059	17.028	0.000
SCORE29	1.015	0.059	17.179	0.000
SCORE30	1.005	0.058	17.368	0.000

F2   WITH

F1	0.005	0.021	0.240	0.810

F3   WITH

F1	-0.032	0.019	-1.707	0.088
F2	0.028	0.019	1.448	0.148

F4   WITH

F1	-0.030	0.020	-1.494	0.135
F2	-0.025	0.021	-1.216	0.224
F3	0.012	0.019	0.650	0.516

Intercepts

SCORE1	-0.003	0.036	-0.072	0.943
SCORE2	-0.020	0.034	-0.576	0.565
SCORE3	0.011	0.036	0.293	0.769
SCORE4	-0.005	0.037	-0.147	0.883
SCORE5	-0.023	0.037	-0.637	0.524
SCORE6	-0.014	0.036	-0.393	0.695
SCORE7	0.009	0.036	0.258	0.796
SCORE8	0.081	0.037	2.207	0.027
SCORE9	0.052	0.034	1.541	0.123
SCORE10	0.082	0.037	2.206	0.027
SCORE11	0.040	0.036	1.109	0.267
SCORE12	0.040	0.035	1.149	0.251
SCORE13	0.007	0.037	0.183	0.854
SCORE14	0.038	0.036	1.075	0.282
SCORE15	0.018	0.037	0.486	0.627
SCORE16	0.040	0.036	1.118	0.264
SCORE17	0.066	0.035	1.863	0.062

SCORE18	0.025	0.035	0.725	0.468
SCORE19	0.002	0.036	0.055	0.956
SCORE20	-0.042	0.035	-1.176	0.240
SCORE21	-0.081	0.036	-2.252	0.024
SCORE22	0.022	0.036	0.615	0.539
SCORE23	-0.041	0.035	-1.164	0.244
SCORE24	-0.062	0.037	-1.688	0.091
SCORE25	-0.003	0.035	-0.091	0.928
SCORE26	0.044	0.035	1.240	0.215
SCORE27	-0.007	0.036	-0.202	0.840
SCORE28	0.016	0.036	0.450	0.653
SCORE29	0.026	0.036	0.724	0.469
SCORE30	0.061	0.035	1.708	0.088

Variances

F1	0.477	0.046	10.261	0.000
F2	0.544	0.048	11.397	0.000
F3	0.419	0.042	9.869	0.000
F4	0.480	0.046	10.456	0.000

Residual Variances

SCORE1	0.535	0.032	16.853	0.000
SCORE2	0.429	0.027	16.126	0.000
SCORE3	0.522	0.031	16.661	0.000
SCORE4	0.481	0.030	15.830	0.000
SCORE5	0.550	0.033	16.720	0.000
SCORE6	0.496	0.030	16.414	0.000
SCORE7	0.494	0.028	17.735	0.000
SCORE8	0.520	0.029	17.885	0.000
SCORE9	0.496	0.027	18.270	0.000
SCORE10	0.566	0.031	18.095	0.000
SCORE11	0.524	0.029	18.062	0.000
SCORE12	0.506	0.028	18.117	0.000
SCORE13	0.535	0.030	17.945	0.000
SCORE14	0.488	0.027	17.825	0.000
SCORE15	0.493	0.028	17.661	0.000
SCORE16	0.545	0.030	18.184	0.000
SCORE17	0.501	0.028	18.023	0.000
SCORE18	0.534	0.031	17.417	0.000
SCORE19	0.500	0.030	16.510	0.000
SCORE20	0.480	0.029	16.603	0.000
SCORE21	0.541	0.032	17.128	0.000
SCORE22	0.534	0.031	17.062	0.000
SCORE23	0.478	0.029	16.576	0.000
SCORE24	0.571	0.033	17.169	0.000
SCORE25	0.503	0.031	16.440	0.000
SCORE26	0.486	0.030	16.249	0.000
SCORE27	0.520	0.032	16.399	0.000
SCORE28	0.527	0.032	16.547	0.000

SCORE29	0.538	0.032	16.564	0.000
SCORE30	0.502	0.031	16.406	0.000

STANDARDIZED MODEL RESULTS

STDYX Standardization

	Estimate	S.E.	Est./S.E.	Two-Tailed P-Value
F1 BY				
SCORE1	0.686	0.022	30.686	0.000
SCORE2	0.725	0.021	35.187	0.000
SCORE3	0.697	0.022	31.862	0.000
SCORE4	0.738	0.020	36.913	0.000
SCORE5	0.695	0.022	31.566	0.000
SCORE6	0.711	0.021	33.450	0.000
F2 BY				
SCORE7	0.724	0.019	38.347	0.000
SCORE8	0.712	0.019	36.509	0.000
SCORE9	0.671	0.021	31.312	0.000
SCORE10	0.690	0.021	33.540	0.000
SCORE11	0.694	0.020	34.020	0.000
SCORE12	0.689	0.021	33.425	0.000
SCORE13	0.706	0.020	35.717	0.000
SCORE14	0.717	0.019	37.318	0.000
SCORE15	0.731	0.019	39.422	0.000
SCORE16	0.681	0.021	32.444	0.000
SCORE17	0.698	0.020	34.681	0.000
F3 BY				
SCORE18	0.663	0.023	28.700	0.000
SCORE19	0.720	0.020	35.166	0.000
SCORE20	0.716	0.021	34.556	0.000
SCORE21	0.683	0.022	30.762	0.000
SCORE22	0.688	0.022	31.243	0.000
SCORE23	0.716	0.021	34.651	0.000
SCORE24	0.680	0.022	30.439	0.000
F4 BY				
SCORE25	0.699	0.022	31.673	0.000
SCORE26	0.709	0.022	32.802	0.000
SCORE27	0.702	0.022	31.997	0.000
SCORE28	0.694	0.022	31.117	0.000
SCORE29	0.692	0.022	30.938	0.000
SCORE30	0.701	0.022	31.940	0.000

```
F2        WITH
   F1                    0.010      0.040       0.240      0.810

F3        WITH
   F1                   -0.071      0.041      -1.725      0.084
   F2                    0.058      0.040       1.460      0.144

F4        WITH
   F1                   -0.063      0.042      -1.507      0.132
   F2                   -0.049      0.040      -1.223      0.221
   F3                    0.027      0.042       0.651      0.515

Intercepts
   SCORE1               -0.003      0.036      -0.072      0.943
   SCORE2               -0.021      0.036      -0.576      0.565
   SCORE3                0.010      0.036       0.293      0.769
   SCORE4               -0.005      0.036      -0.147      0.883
   SCORE5               -0.023      0.036      -0.637      0.524
   SCORE6               -0.014      0.036      -0.393      0.695
   SCORE7                0.009      0.036       0.258      0.796
   SCORE8                0.079      0.036       2.203      0.028
   SCORE9                0.055      0.036       1.540      0.123
   SCORE10               0.079      0.036       2.203      0.028
   SCORE11               0.040      0.036       1.109      0.268
   SCORE12               0.041      0.036       1.149      0.251
   SCORE13               0.007      0.036       0.183      0.854
   SCORE14               0.038      0.036       1.075      0.283
   SCORE15               0.017      0.036       0.486      0.627
   SCORE16               0.040      0.036       1.118      0.264
   SCORE17               0.067      0.036       1.861      0.063
   SCORE18               0.026      0.036       0.725      0.468
   SCORE19               0.002      0.036       0.055      0.956
   SCORE20              -0.042      0.036      -1.176      0.240
   SCORE21              -0.080      0.036      -2.248      0.025
   SCORE22               0.022      0.036       0.615      0.539
   SCORE23              -0.042      0.036      -1.163      0.245
   SCORE24              -0.060      0.036      -1.686      0.092
   SCORE25              -0.003      0.036      -0.091      0.928
   SCORE26               0.044      0.036       1.240      0.215
   SCORE27              -0.007      0.036      -0.202      0.840
   SCORE28               0.016      0.036       0.450      0.653
   SCORE29               0.026      0.036       0.724      0.469
   SCORE30               0.061      0.036       1.706      0.088

Variances
   F1                    1.000      0.000     999.000    999.000
   F2                    1.000      0.000     999.000    999.000
   F3                    1.000      0.000     999.000    999.000
   F4                    1.000      0.000     999.000    999.000
```

```
Residual Variances
    SCORE1          0.529       0.031       17.216      0.000
    SCORE2          0.474       0.030       15.854      0.000
    SCORE3          0.514       0.031       16.825      0.000
    SCORE4          0.455       0.030       15.391      0.000
    SCORE5          0.518       0.031       16.934      0.000
    SCORE6          0.494       0.030       16.351      0.000
    SCORE7          0.476       0.027       17.399      0.000
    SCORE8          0.494       0.028       17.792      0.000
    SCORE9          0.549       0.029       19.091      0.000
    SCORE10         0.524       0.028       18.471      0.000
    SCORE11         0.519       0.028       18.355      0.000
    SCORE12         0.526       0.028       18.518      0.000
    SCORE13         0.502       0.028       17.968      0.000
    SCORE14         0.486       0.028       17.622      0.000
    SCORE15         0.466       0.027       17.202      0.000
    SCORE16         0.536       0.029       18.769      0.000
    SCORE17         0.512       0.028       18.208      0.000
    SCORE18         0.560       0.031       18.293      0.000
    SCORE19         0.481       0.030       16.296      0.000
    SCORE20         0.488       0.030       16.463      0.000
    SCORE21         0.533       0.030       17.570      0.000
    SCORE22         0.527       0.030       17.417      0.000
    SCORE23         0.487       0.030       16.423      0.000
    SCORE24         0.537       0.030       17.672      0.000
    SCORE25         0.511       0.031       16.570      0.000
    SCORE26         0.497       0.031       16.219      0.000
    SCORE27         0.507       0.031       16.483      0.000
    SCORE28         0.519       0.031       16.768      0.000
    SCORE29         0.521       0.031       16.811      0.000
    SCORE30         0.508       0.031       16.499      0.000
```

It is possible to specify the use of an estimator other than ML, simply by including the estimator= subcommand under analysis. A wide variety of options are available, and the interested reader is encouraged to investigate these.

Structural Equation Models

Researchers are frequently interested in investigating relationships among constructs, such as depression and anxiety, or academic achievement and motivation. When using observed variables, these types of questions are typically addressed by using regression or some similar approach. There exist analogous methods, known collectively as structural equation models (SEMs), for researchers working with latent variables. Thus, the same types of questions that can be addressed by using regression with observed variables can also be investigated for latent variables. Consider a researcher investigating relationships among five latent variables. Each of the factors is measured by multiple continuous observed indicators, and the model that is proposed appears in Figure 9.2.

This model suggests that the latent variables f1, f2, and f3 are related to the latent variable f5 both directly and as mediated by latent variable f4. In addition, it is hypothesized that f1, f2, and f3 are correlated with one another. Each of the latent variables f1 through f4 has three observed indicators, and f5 has eight indicators. Equation 9.11 expresses the basic SEM, whereby η represents one or more endogenous (i.e., dependent) latent variables and γ represents one or more exogenous (i.e., independent) latent variables.

$$\eta = B\gamma + \zeta \tag{9.11}$$

The relationship between a pair of factors is expressed through B, which can be viewed as very similar to a regression coefficient. In addition, the model includes random error, ζ, with a mean of 0 and variance of ϕ.

The relationships in Equation 9.11 represent the structural part of the model, whereas the relationships between the observed indicators and latent variables represent the measurement part of the model. The measurement portion of an SEM simply consists of the individual CFA models appropriate for each

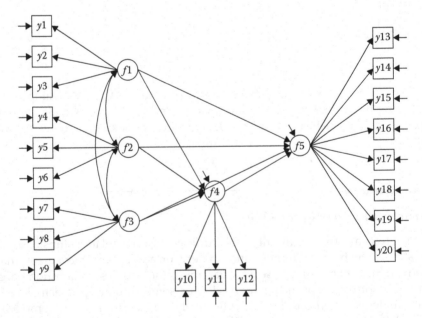

FIGURE 9.2
Structural equation model for partially mediated model.

of the factors in the structural portion. The fitting of an SEM occurs in two steps, with the first involving estimation of CFA models for each of the latent factors. If the measurement model does not yield good data fit, then the structural part of the model cannot be estimated, because the latent variables cannot be assumed to exist as hypothesized. However, by assuming good fit for the measurement models for each factor, the structural portion of the model can be fit to the data. The focus of the SEM is usually on the model parameters that assess the hypothesized relationships among the factors, much in the way that the focus of regression modeling is on the slopes. In addition, in the context of SEM, we are interested in ascertaining model fit, which is done by using the same tools that we used with CFA. Finally, often when using SEM, we will have multiple models that are to be compared with one another. These comparisons can be made by using the χ^2 difference test for nested models, as well as AIC, BIC, and aBIC for both nested and non-nested models.

Fitting SEM by Using Mplus

The Mplus commands to fit the model in Figure 9.2 appear below and are contained in the file model_9.3.inp.

```
TITLE: SEM model

DATA: FILE IS model9_3.dat;

VARIABLE:   NAMES ARE y1-y20;
analysis:   bootstrap=1000;
MODEL:
        f1 by y1@1 y2-y3;
        f2 by y4@1 y5-y6;
        f3 by y7@1 y8-y9;
        f4 by y10@1 y11-y12;
        f5 by y13@1 y14-y20;

        y1-y20;
        f1-f5;

    f4 on f1 f2 f3;
    f5 on f1 f2 f3 f4;

MODEL INDIRECT:
    f5 ind f4 f1;
    f5 ind f4 f2;
    f5 ind f4 f3;

output:   standardized cinterval;
```

The measurement models are specified first, and the factor loading for the first observed variable in the list associated with each factor is set to 1 by using @1. Estimates of the variances for the observed and latent variables are then requested y1-y20; f1-f5;. The structural portion of the model is specified next, including the request for modeling of the indirect effects relating f1, f2, and f2–f5, through the mediator f4. The standardized model parameter estimates are requested in the output command, along with the confidence intervals for the parameter estimates, which will be based on the 1000 bootstrap samples requested in the program. Relevant portions of the resulting output appear below.

We can first assess the fit of the model, using the same statistical tools that were employed for assessing fit of CFA models.

```
THE MODEL ESTIMATION TERMINATED NORMALLY


MODEL FIT INFORMATION

Number of Free Parameters                        70

Loglikelihood

          H0 Value                       -25673.160
          H1 Value                       -25592.513

Information Criteria

          Akaike (AIC)                    51486.320
          Bayesian (BIC)                  51814.505
          Sample-Size Adjusted BIC        51592.215
            (n* = (n + 2) / 24)

Chi-Square Test of Model Fit

          Value                             161.293
          Degrees of Freedom                    160
          P-Value                            0.4565

RMSEA (Root Mean Square Error Of Approximation)

          Estimate                            0.003
          90 Percent C.I.                     0.000    0.017
          Probability RMSEA <= .05            1.000

CFI/TLI

          CFI                                 1.000
          TLI                                 1.000
```

```
Chi-Square Test of Model Fit for the Baseline Model

            Value                          10888.178
            Degrees of Freedom                   190
            P-Value                           0.0000

SRMR (Standardized Root Mean Square Residual)

            Value                              0.016
```

The guidelines for assessing model quality based on the various fit indices that were discussed previously in the context of CFA are also applicable when considering the fit of an SEM. All of these indices, including the χ^2 goodness of fit test, suggest that the model fits the data well. The unstandardized structural model parameter estimates appear below.

```
F4        ON
    F1                    0.732      0.060     12.115      0.000
    F2                    0.904      0.071     12.770      0.000
    F3                    0.402      0.117      3.433      0.001

F5        ON
    F1                    0.247      0.088      2.800      0.005
    F2                    0.442      0.103      4.283      0.000
    F3                    0.255      0.103      2.467      0.014
    F4                    0.755      0.086      8.765      0.000

F2        WITH
    F1                    0.084      0.054      1.533      0.125

F3        WITH
    F1                    0.026      0.039      0.661      0.509
    F2                    0.084      0.041      2.056      0.040
```

These results indicate that there were statistically significant direct relationships of F1, F2, and F3 with F4, and separate relationships of F1, F2, F3, and F4 with F5. In addition, only F2 and F3 exhibited a statistically significant relationship with one another. Although they are not displayed here, the standardized estimates are also included in the output, as are the bootstrap confidence intervals for all of the model parameter estimates. Of particular interest in this case are the confidence intervals for the indirect effects, because the usual standard error estimates that are obtained for the mediation coefficients are not accurate and the bootstrap approach to constructing confidence intervals for these effects has been found to work well in many situations (Preacher & Hayes, 2008). The bootstrap confidence intervals for the indirect effects appear in the following.

CONFIDENCE INTERVALS OF TOTAL, TOTAL INDIRECT, SPECIFIC
INDIRECT, AND DIRECT EFFECTS

	Lower .5%	Lower 2.5%	Lower 5%	Estimate	Upper 5%	Upper 2.5%	Upper .5%
Effects from F1 to F5							
Indirect	0.350	0.399	0.423	0.553	0.682	0.707	0.755
Effects from F2 to F5							
Indirect	0.460	0.513	0.541	0.682	0.824	0.851	0.904
Effects from F3 to F5							
Indirect	0.060	0.118	0.148	0.303	0.459	0.488	0.547

The indirect relationship between f1 and f5 through f4 was estimated to be
0.533. Given that the 95% bootstrap confidence interval (lower 2.5% and upper
2.5%) did not include 0, we can conclude that the indirect effect was indeed
statistically significant. In other words, there was an indirect relationship
between f1 and f5 through f4. Similar results were evident for the indirect
effects of f2 and f3 as well.

Growth Curve Models

In the previous sections of this chapter, we have described latent variable
modeling in terms of understanding the nature of the latent variables in
terms of the observed indicators (factor analysis), and how latent variables
can be related to one another by using models similar in form to regression
equations. Next, we consider the application of latent variable modeling to
the analysis of change in the observed variables over time. In this context,
a researcher may have collected data on a sample at multiple points in time
and would like to know whether and how the scores on these measures
change over time. The researcher interested in assessing change over time
in a variable has at her disposal a number of statistical tools. For example,
she could use the repeated measures analysis of variance (ANOVA) to compare means of a single measurement taken at multiple points in time or the
repeated measures multivariate analysis of variance (MANOVA) for multiple dependent variables. Furthermore, as we saw in Chapter 6, multilevel
models can also be used for modeling longitudinal data. Another option for
researchers interested in modeling longitudinal data is growth curve modeling (GCM), sometimes also referred to as latent growth curve modeling.
With GCM, we use latent variable modeling to estimate growth over time
and the mean starting value of the time series. In this way, we will obtain

estimates for the average growth rate between adjacent times, as well as the relationship between this growth rate and the initial value of the series. Because GCM is a latent variable technique, it also allows the researcher to build structural models, linking change over time in multiple variables simultaneously and linking such change to observed variables (e.g., gender and ethnicity).

As noted above, basic GCMs involve two latent variables, the initial status of the construct being measured and the change in that construct over time. The GCM appears in Figure 9.3.

In the basic linear GCM, the intercept (I) represents initial status of the outcome variable of interest and the slope (S) represents the rate of change in the variable. In this example, there are four observed variables, labeled score1 through score4. The loadings that relate these four observed variables to the latent traits are constrained to be specific values reflecting the nature of the growth being modeled. For example, in order to model linear growth, we would set the loadings for S to be 0, 1, 2, and 3, such that the equal distance between values reflects the linear growth trajectory. With regard to the intercept, the coefficients are all set to 1, and this will be the case regardless of the hypothesis regarding change over time. By setting the coefficient for the initial time point to 0, its mean is the estimate for the mean of I; that is, I reflects the mean starting value for the variable of interest. The GCM will also provide an estimate of the covariance (or correlation) between the change over time (S) and the initial mean value (I). Finally, we are not limited to fitting only linear change trajectories with GCMs. For example, if we believe that the change in scores follows a quadratic trend over time, we can fit such a model and test whether the quadratic change term is statistically significant. It is possible to fit other nonlinear change functions as well, using GCMs. As we will see, fitting GCMs by using Mplus is very straightforward.

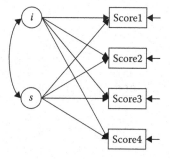

FIGURE 9.3
Linear growth curve model.

Fitting GCM by Using Mplus

For this example, we have a standardized mathematics achievement test that has been measured at four points in time for 379 individuals. The data are contained in the file gcm.dat. We will begin by fitting a linear GCM. The Mplus program to do so appears below.

```
TITLE:     Growth Curve Model
DATA:      FILE IS gcm.dat;
VARIABLE: NAMES ARE score1-score4;
MODEL:     i s | score1@0 score2@1 score3@2 score4@3;
output: standardized;
plot:     type=plot3; series=score1-score4(*);
```

The portions of this program that are different from what we have examined earlier are in the MODEL and PLOT statements. In order to fit the linear GCM, we need to specify the intercept and slope values (i and s) as being a function of the four observed measures. In addition, by using @, we are able to set the loadings for each score to be a specific value, for example, 0, 1, 2, and 3. With PLOT, we will be able to graph the means of the four scores over time and thereby gain a sense for the nature of the growth that occurred. Because the scores are all to be used in the same plot, we must specify the series= command. Including (*) indicates that the x-axis of the graph will start at 0 and be incremented by 1 for each of the scores; that is, it will go from 0 to 3 in this case. The relevant output appears below.

MODEL RESULTS

		Estimate	S.E.	Est./S.E.	Two-Tailed P-Value
I					
	SCORE1	1.000	0.000	999.000	999.000
	SCORE2	1.000	0.000	999.000	999.000
	SCORE3	1.000	0.000	999.000	999.000
	SCORE4	1.000	0.000	999.000	999.000
S					
	SCORE1	0.000	0.000	999.000	999.000
	SCORE2	1.000	0.000	999.000	999.000
	SCORE3	2.000	0.000	999.000	999.000
	SCORE4	3.000	0.000	999.000	999.000
S	WITH				
	I	0.024	0.105	0.229	0.818
Means					
	I	0.333	0.078	4.260	0.000
	S	2.135	0.077	27.670	0.000

```
Intercepts
   SCORE1              0.000       0.000     999.000     999.000
   SCORE2              0.000       0.000     999.000     999.000
   SCORE3              0.000       0.000     999.000     999.000
   SCORE4              0.000       0.000     999.000     999.000

Variances
   I                   0.901       0.205       4.400       0.000
   S                   1.778       0.157      11.356       0.000

Residual Variances
   SCORE1              0.969       0.210       4.623       0.000
   SCORE2              1.641       0.142      11.554       0.000
   SCORE3             -0.015       0.155      -0.095       0.924
   SCORE4              7.270       0.590      12.312       0.000


STANDARDIZED MODEL RESULTS


STDYX Standardization

S        WITH
   I                   0.019       0.085       0.226       0.821

Means
   I                   0.351       0.077       4.582       0.000
   S                   1.601       0.095      16.864       0.000
```

The output above includes the full set of unstandardized results, as well as the relevant standardized values. Note that with GCMs, the loadings are not of interest, because they have been set to specific values in order to identify and fit the model. In this case, we will focus on the means for I and S, as well as on the correlation between the two. The average starting value for the test scores was 0.333 (unstandardized mean of I), and the mean change over time was 2.135. In other words, the average rate of growth between adjacent time points was 2.135. Both of these values were statistically significant. Thus, we can conclude that there was significant change in the score values over time. The covariance between the two variables was not statistically significant, which tells us that change over time was not related to the initial value of the score. The advantage of referring to the standardized values in this case is that we have an estimate of the correlation coefficient, which in this case is 0.019, further indicating the lack of a relationship between the initial score and the rate of change in score values. The graph plotting the observed means appears in Figure 9.4. We can access this tool simply by clicking on ☑ in the menu bar and then selecting the graph that we would like to see.

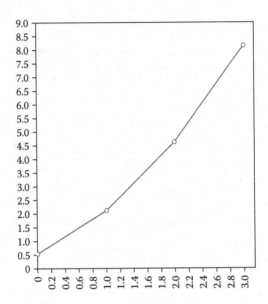

FIGURE 9.4
Plot of score mean by time.

From this figure, we can see that there is a positive change in score means over time and that this change may not be totally linear in nature. Thus, it may behoove us to fit a quadratic model to the data.

If we believe that there may be a quadratic change over time, it is very easy to fit the appropriate model in Mplus. We simply replace the MODEL command for the linear GCM with the following:

```
MODEL: i s q | score1@0 score2@1 score3@2 score4@3;
```

By adding in q, we are requesting Mplus to fit the quadratic term in the model. The relevant output appears below.

```
MODEL RESULTS

S           WITH
    I                   0.030      0.352      0.085      0.933

Q           WITH
    I                   0.073      0.091      0.804      0.421
    S                  -0.268      0.132     -2.035      0.042

Means
    I                   0.538      0.070      7.645      0.000
    S                   1.068      0.088     12.151      0.000
    Q                   0.490      0.036     13.641      0.000
```

```
STANDARDIZED MODEL RESULTS


STDYX Standardization
S         WITH
    I                   0.030       0.370       0.082       0.934

Q         WITH
    I                   0.119       0.132       0.904       0.366
    S                  -0.472       0.116      -4.050       0.000

Means
    I                   0.524       0.105       4.996       0.000
    S                   1.123       0.273       4.115       0.000
    Q                   0.820       0.099       8.297       0.000
```

From these results, we can conclude that there was indeed a significant quadratic effect and that the quadratic term was negatively correlated with the linear term ($r = -0.472$). This means that individuals who experienced more linear growth had less quadratic change over time in their scores. Using the information indices (i.e., AIC, BIC, and aBIC) that we have discussed before, we can decide which model should be retained. The values for each of the information indices were smaller for the quadratic model, suggesting that it provides better fit to the data than the linear one.

There are a number of ways in which the GCM paradigm can be extended. For example, change over time can be modeled for multiple variables simultaneously (e.g., reading, math, and science achievement) in the same model, with the relationships between these change functions correlated with one another. In addition, observed and latent variables can serve as predictors of the slope and intercept variables. Thus, a researcher could investigate whether the change in scores over time differed for males and females, as an example. All of these variations on the GCM are available in Mplus, and the interested reader is encouraged to investigate them further.

Item Response Theory

Item response theory (IRT) is the latent variable modeling framework that is used by researchers working with responses to items on cognitive and achievement tests, measures of mood and personality, and other similar types of instruments. IRT models link observed item responses with the latent construct that is being measured by the instrument and provide their users with estimates of both the individual examinee levels of the latent construct and the information about the performance of the individual items on the scale. There are a variety of such models for use with both dichotomous and polytomous items and that provide varying amounts of information about the items. All of these IRT models yield a parameter estimate specific to the respondents (person parameter), which is simply the estimate of the latent trait being measured by the scale (e.g., reading ability, depression, and extraversion) and which is represented as θ. The item parameters include location on the latent trait scale (*b*), ability to differentiate among individuals with different levels of the construct (*a*), and, in some cases, the likelihood that an individual will endorse the item due to chance (*c*). Polytomous IRT models are characterized by multiple item location parameters, known as thresholds (τ), which correspond to intersections of different response options on the latent trait scale. Following is a brief description of some basic dichotomous and polytomous IRT models. There is not sufficient space in this chapter to discuss all of the intricacies of these models. However, we do endeavor to present some basic information, on which the interested readers can build as they wish.

The most general dichotomous IRT model is the 3-parameter logistic (3PL), which can be written as

$$P\left(x_j = 1 \mid \theta_i, a_j, b_j, c_j\right) = c_i + \left(1 - c_i\right)\frac{e^{a_j\left(\theta_i - b_j\right)}}{1 + e^{a_j\left(\theta_i - b_j\right)}} \tag{9.12}$$

where:
  θ_i is the latent trait being measured for subject *i*
  b_j is the difficulty (location) for item *j*
  a_j is the discrimination for item *j*
  c_i is the pseudo-chance parameter for item *j*

The 3PL model is most appropriate for use with item where guessing a correct item response is a possibility. When such is not the case, such as for a scale designed to assess depressive symptoms, the 2-parameter logistic (2PL) model may be more appropriate. It is similar in form to the 3PL model, except that the pseudo-chance parameter is set to 0.

$$P\left(x_j = 1 \mid \theta, a_j, b_j\right) = \frac{e^{a_j\left(\theta_i - b_j\right)}}{1 + e^{a_j\left(\theta_i, -b_j\right)}} \tag{9.13}$$

In some cases, the researcher may want to estimate a model for which only a single discrimination parameter is estimated for each item (1PL) or where the item discrimination is set to 1 for all of the items (Rasch). Such models may be particularly useful when working with smaller samples, where parameter estimation of the more complex models may be compromised. The 1PL model takes the form:

$$P\left(x_j = 1 \mid \theta_i, a_j, b_j\right) = \frac{e^{a_j\left(\theta_i, -b_j\right)}}{1 + e^{a_j\left(\theta_i - b_j\right)}} \tag{9.14}$$

In this model, a single-item discrimination parameter is estimated for all of the items. The Rasch model sets $a = 1$ for all of the items.

$$P\left(x_j = 1 \mid \theta_i, b_j\right) = \frac{e^{\left(\theta_i - b_j\right)}}{1 + e^{\left(\theta_i - b_j\right)}} \tag{9.15}$$

There also exist IRT models for use with polytomous items; that is, those with more than two categories. These models are similar in form to those for dichotomous items, with the primary difference being in the manner in which item location is expressed. In addition, these models do not incorporate the pseudo-guessing parameter values, so that the most complex of them is equivalent to the 2PL model for dichotomous items. One of the simplest of these polytomous item models is the partial credit model (PCM), which is expressed as

$$p\left(x_j = k \mid \theta_i, \delta_{jh}\right) = \frac{e^{\sum_{h=0}^{x_j}\left(\theta_i - \delta_{jh}\right)}}{e^{\sum_{h=0}^{m_j}\left(\theta_i - \delta_{jh}\right)}} \tag{9.16}$$

where:
x_j is the response x to item j
θ_i is the level of the latent trait being measured for subject i
δ_{jh} is the threshold parameter h for item j
m_j is the total number of possible response categories for item j

The θ parameter has the same meaning for polytomous item models as it has for dichotomous items; that is, it is a measure of the latent trait being measured by the scale. The δ_{jh} parameter measures the location of a specific item response option rather than the item itself. Put another way, δ_{jh} reflects the relative likelihood that an individual will select item response option h versus $h-1$. For an item with m possible response options, there are $m-1$ thresholds. Larger values of δ_{jh} indicate that a respondent would need to have a higher value of θ in order to select response h, as opposed to $h-1$. The PCM in Equation 9.15 is a Rasch model for polytomous data, because the item discrimination parameter (a) is set equal to 1 for all items. It is also possible to estimate a common a for all of the items, rather than setting it to 1, thus leading to the 1PL version of the PCM.

By including a unique discrimination estimate for each item, the PCM can be extended to the generalized partial credit model (GPCM).

$$p\left(x_j = k \mid \theta_i, a_j, \delta_{jh}\right) = \frac{e^{\sum_{h=0}^{x_j} a_j\left(\theta_i - \delta_{jh}\right)}}{e^{\sum_{h=0}^{m_j} a_j\left(\theta_i - \delta_{jh}\right)}} \tag{9.17}$$

An alternative model for polytomous data that incorporates individual discrimination parameter values for each item is the graded response model

(GRM), developed by Samejima (1969). Although like the GPCM, the GRM provides unique discrimination parameter estimates for each item, the two models differ with respect to the ways in which they conceptualize transition from one response option to another. The GPCM treats these transitions as occurring in a binary fashion, whereby the probability of an individual selecting between two adjacent categories (e.g., 1 versus 2) is reflected in the threshold values. The GRM, in contrast, expresses the comparison of item response choices in terms of the likelihood of an individual selecting a particular response option or one higher, versus responding with a lower option. In this framework, a response of 2 or higher is modeled against a response of 1, for example. The GRM takes the form:

$$p\left(x_j = k \mid \theta_i, a, \delta_{jh}\right) = \frac{e^{a_j\left(\theta_i - \delta_{jh}\right)}}{1 + e^{a_j\left(\theta_i - \delta_{jh}\right)}} \tag{9.18}$$

Fitting IRT Models by Using Mplus

In order to demonstrate the fitting of IRT models by using Mplus, let us consider an example dataset consisting of dichotomous responses to 25 items designed to measure anxiety symptoms. The instrument was completed by 417 individuals. Higher values of θ indicate a greater sense of anxiety on the part of the respondent. The Mplus program to fit the Rasch model appears below.

```
TITLE: Rasch IRT model
DATA:   FILE IS anxiety.csv;
VARIABLE:     NAMES ARE anxq1 anxq2 anxq3 anxq4 anxq5 anxq6
anxq7 anxq8 anxq9 anxq10
anxq11 anxq12 anxq13 anxq14 anxq15 anxq16 anxq17 anxq18 anxq19
anxq20 anxq21 anxq22 anxq23 anxq24 anxq25;

CATEGORICAL ARE anxq1-anxq25;

ANALYSIS:     ESTIMATOR = MLR;

MODEL: anx BY anxq1@1 anxq2-anxq25@1;
       anx@1;
       [anx@0];

PLOT:  TYPE = PLOT3;

output: tech5 tech10;
```

Much of this program resembles earlier ones; however, there are some unique features called upon when fitting IRT models. First, note that we must be sure to indicate that the item response variables are categorical in nature. We also must indicate that the estimator to be used is maximum likelihood with robust standard errors, based on the sandwich estimator (MLR). In the MODEL statement, we link the items (anxq1-anxq25) to the latent trait, which we call anx, and set the mean and variance of the latent trait to 0 and 1, respectively. We have specified a Rasch model by setting the discrimination parameter values to be equal to 1 for all of the items by using @1. The item difficulty parameters will still be freely estimated. The model fit output appears below. As with many other applications in this book, we will use the information criteria to compare the fit of different IRT models, in order to select the one that appears to be optimal.

```
MODEL FIT INFORMATION

Number of Free Parameters                        25

Loglikelihood

        H0 Value                          -3921.831
        H0 Scaling Correction Factor         0.7694
           for MLR

Information Criteria

        Akaike (AIC)                       7893.662
        Bayesian (BIC)                     7994.489
        Sample-Size Adjusted BIC           7915.157
           (n* = (n + 2) / 24)
```

The IRT model parameter estimates appear below.

```
IRT PARAMETERIZATION

  Item Discriminations


ANX        BY
    ANXQ1                1.000        0.000        0.000        1.000
    ANXQ2                1.000        0.000        0.000        1.000
    ANXQ3                1.000        0.000        0.000        1.000
    ANXQ4                1.000        0.000        0.000        1.000
    ANXQ5                1.000        0.000        0.000        1.000
    ANXQ6                1.000        0.000        0.000        1.000
    ANXQ7                1.000        0.000        0.000        1.000
    ANXQ8                1.000        0.000        0.000        1.000
    ANXQ9                1.000        0.000        0.000        1.000
    ANXQ10               1.000        0.000        0.000        1.000
    ANXQ11               1.000        0.000        0.000        1.000
    ANXQ12               1.000        0.000        0.000        1.000
    ANXQ13               1.000        0.000        0.000        1.000
    ANXQ14               1.000        0.000        0.000        1.000
    ANXQ15               1.000        0.000        0.000        1.000
    ANXQ16               1.000        0.000        0.000        1.000
    ANXQ17               1.000        0.000        0.000        1.000
```

ANXQ18	1.000	0.000	0.000	1.000
ANXQ19	1.000	0.000	0.000	1.000
ANXQ20	1.000	0.000	0.000	1.000
ANXQ21	1.000	0.000	0.000	1.000
ANXQ22	1.000	0.000	0.000	1.000
ANXQ23	1.000	0.000	0.000	1.000
ANXQ24	1.000	0.000	0.000	1.000
ANXQ25	1.000	0.000	0.000	1.000

Means
ANX	0.000	0.000	0.000	1.000

Item Difficulties
ANXQ1	-0.872	0.163	-5.364	0.000
ANXQ2	-0.413	0.162	-2.544	0.011
ANXQ3	0.020	0.165	0.121	0.904
ANXQ4	0.732	0.164	4.470	0.000
ANXQ5	0.618	0.166	3.717	0.000
ANXQ6	1.280	0.166	7.695	0.000
ANXQ7	-0.234	0.169	-1.388	0.165
ANXQ8	-1.382	0.163	-8.456	0.000
ANXQ9	-0.306	0.164	-1.862	0.063
ANXQ10	0.020	0.166	0.120	0.904
ANXQ11	0.334	0.165	2.027	0.043
ANXQ12	0.447	0.165	2.712	0.007
ANXQ13	0.848	0.164	5.157	0.000
ANXQ14	0.020	0.170	0.118	0.906
ANXQ15	1.043	0.164	6.362	0.000
ANXQ16	-1.188	0.165	-7.199	0.000
ANXQ17	-0.679	0.162	-4.181	0.000
ANXQ18	-1.436	0.168	-8.553	0.000
ANXQ19	-0.053	0.166	-0.320	0.749
ANXQ20	0.409	0.164	2.499	0.012
ANXQ21	0.868	0.164	5.279	0.000
ANXQ22	0.409	0.167	2.445	0.014
ANXQ23	0.561	0.167	3.358	0.001
ANXQ24	0.057	0.165	0.342	0.732
ANXQ25	0.984	0.164	5.997	0.000

Variances
ANX	1.000	0.000	0.000	1.000

Because this is a Rasch model, the item discrimination values have all been set to 1. From the item difficulty estimates, we can see that item 6 is the most difficult for respondents to endorse ($b = 1.280$) and item 18 is the easiest to endorse ($b = -1.436$). In addition, there appears to be a mix of item difficulty values, with no overall pattern toward a more difficult or easier scale overall.

By using the plotting capabilities in Mplus, we can visualize the item characteristic curve (ICC) for each of the items. For dichotomous items, the ICC displays the relationship between the probability of endorsing the item (or a correct item response) and the value on the latent trait. In order to obtain the ICC, we click on 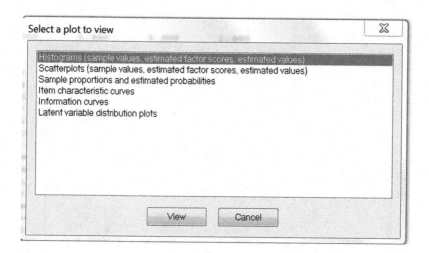 to obtain the following window.

Select a plot to view

Histograms (sample values, estimated factor scores, estimated values)
Scatterplots (sample values, estimated factor scores, estimated values)
Sample proportions and estimated probabilities
Item characteristic curves
Information curves
Latent variable distribution plots

View   Cancel

We then select Item characteristic curves and click View.

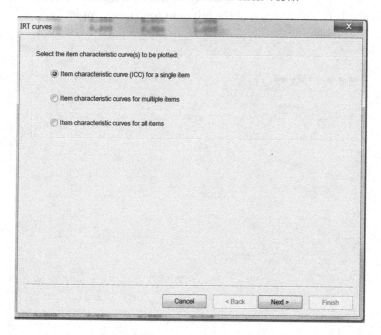

IRT curves

Select the item characteristic curve(s) to be plotted:

◉ Item characteristic curve (ICC) for a single item

○ Item characteristic curves for multiple items

○ Item characteristic curves for all items

Cancel   < Back   Next >   Finish

We then have the option to view an ICC for one item, several items, or all of the items. For a fairly large scale, such as this one, it might be most beneficial to visualize ICCs for individual items. In order to do so, we would click Next.

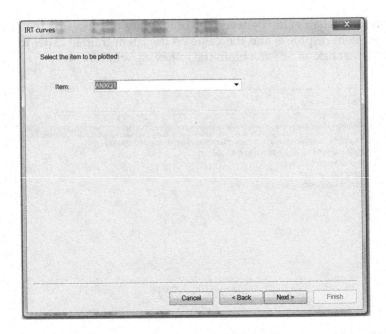

From this pull-down menu, we can select the item for which we would like to view the ICC. We will select the 8th item for this example.

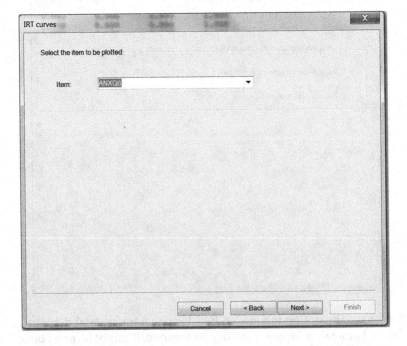

We would then click Next.

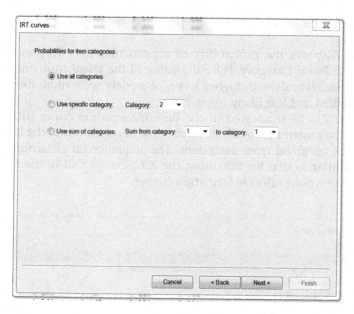

We now have the choice of using all categories (Yes and No, in this example), or we can select only one of the categories (e.g., Yes). For this example, we will use both categories. After clicking Next, we obtain the following table, from which we select the variable for the *x*-axis (by default, it is the latent trait being measured) and the range of values for the *x*-axis. In this case, it is +/−3 standard deviations of the latent trait. Assuming that we are satisfied with the default settings, we can click Next.

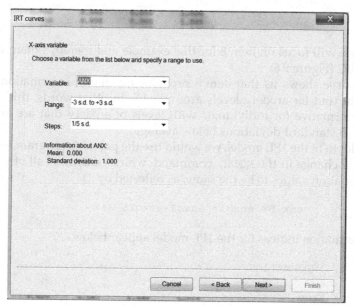

We can click Finish in the next window and obtain the following figure (Figure 9.5).

This ICC displays the probability of a particular item response (Yes = Category 2, No = Category 1) by the value of the latent trait, anxiety. It is clear that individuals with higher levels of anxiety were more likely to say Yes to the item and less likely to say No.

We may also be interested in the item information curve (IIC), which indicates the amount of information about different levels of the latent trait that can be obtained from each item. The sequence for obtaining the IIC is very similar to that for obtaining the ICC, except that in the following window, we would select Information curves.

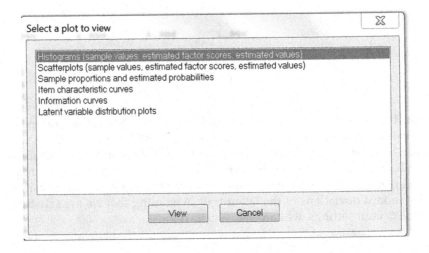

Again, we will focus on item 8 for this example and thereby obtain the following IIC (Figure 9.6).

This curve shows us that item 8 provides maximum information about the latent trait for anxiety levels around −1.5. In other words, this item is more informative for individuals with levels of anxiety that are approximately 1.5 standard deviations below average.

In order to fit the 1PL model, we would use the previous program, with the following change in the MODEL command, which constrains all of the item discrimination values to be the same, as reflected by (1).

```
MODEL:        anx BY anxq1* anxq2-anxq25 (1);
```

The information indices for the 1PL model appear below.

```
MODEL FIT INFORMATION

Number of Free Parameters                    26
```

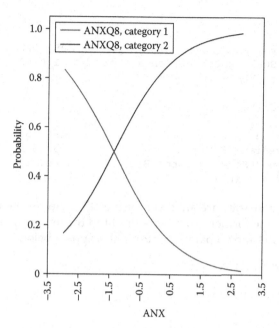

FIGURE 9.5
ICC for anxiety scale item 8, Rasch model.

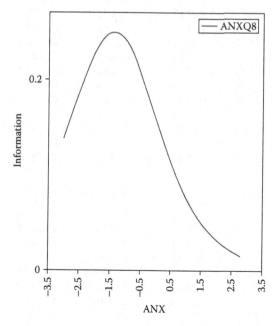

FIGURE 9.6
IIC for anxiety scale item 8, Rasch model.

```
Loglikelihood

        H0 Value                          -3053.328
        H0 Scaling Correction Factor        1.0222
           for MLR

Information Criteria

        Akaike (AIC)                       6158.655
        Bayesian (BIC)                     6263.515
        Sample-Size Adjusted BIC           6181.010
           (n* = (n + 2) / 24)
```

The AIC, BIC, and aBIC are all lower for the 1PL than for the Rasch model, indicating that the former provides better fit to the item response data than the latter. The IRT model parameter estimates appear below.

```
IRT PARAMETERIZATION

  Item Discriminations

ANX        BY
    ANXQ1             5.371      0.384     13.977      0.000
    ANXQ2             5.371      0.384     13.977      0.000
    ANXQ3             5.371      0.384     13.977      0.000
    ANXQ4             5.371      0.384     13.977      0.000
    ANXQ5             5.371      0.384     13.977      0.000
    ANXQ6             5.371      0.384     13.977      0.000
    ANXQ7             5.371      0.384     13.977      0.000
    ANXQ8             5.371      0.384     13.977      0.000
    ANXQ9             5.371      0.384     13.977      0.000
    ANXQ10            5.371      0.384     13.977      0.000
    ANXQ11            5.371      0.384     13.977      0.000
    ANXQ12            5.371      0.384     13.977      0.000
    ANXQ13            5.371      0.384     13.977      0.000
    ANXQ14            5.371      0.384     13.977      0.000
    ANXQ15            5.371      0.384     13.977      0.000
    ANXQ16            5.371      0.384     13.977      0.000
    ANXQ17            5.371      0.384     13.977      0.000
    ANXQ18            5.371      0.384     13.977      0.000
    ANXQ19            5.371      0.384     13.977      0.000
    ANXQ20            5.371      0.384     13.977      0.000
    ANXQ21            5.371      0.384     13.977      0.000
    ANXQ22            5.371      0.384     13.977      0.000
    ANXQ23            5.371      0.384     13.977      0.000
    ANXQ24            5.371      0.384     13.977      0.000
    ANXQ25            5.371      0.384     13.977      0.000

Means
    ANX               0.000      0.000      0.000      1.000
```

```
Item Difficulties
```

ANXQ1	-0.356	0.068	-5.259	0.000
ANXQ2	-0.216	0.069	-3.137	0.002
ANXQ3	-0.082	0.073	-1.136	0.256
ANXQ4	0.158	0.081	1.953	0.051
ANXQ5	0.116	0.080	1.452	0.147
ANXQ6	0.381	0.092	4.123	0.000
ANXQ7	-0.161	0.072	-2.233	0.026
ANXQ8	-0.515	0.068	-7.551	0.000
ANXQ9	-0.183	0.070	-2.612	0.009
ANXQ10	-0.082	0.073	-1.129	0.259
ANXQ11	0.019	0.076	0.249	0.803
ANXQ12	0.057	0.077	0.736	0.461
ANXQ13	0.201	0.083	2.434	0.015
ANXQ14	-0.082	0.074	-1.107	0.268
ANXQ15	0.279	0.086	3.240	0.001
ANXQ16	-0.453	0.068	-6.634	0.000
ANXQ17	-0.297	0.068	-4.372	0.000
ANXQ18	-0.532	0.070	-7.650	0.000
ANXQ19	-0.105	0.072	-1.453	0.146
ANXQ20	0.044	0.076	0.577	0.564
ANXQ21	0.209	0.083	2.516	0.012
ANXQ22	0.044	0.077	0.567	0.571
ANXQ23	0.096	0.079	1.208	0.227
ANXQ24	-0.071	0.073	-0.970	0.332
ANXQ25	0.255	0.085	3.000	0.003

```
Variances
```

ANX	1.000	0.000	0.000	1.000

Perhaps the most notable difference in the results for the 1PL and Rasch models is in the value of the discrimination parameter estimate. Whereas for the Rasch model, it was set equal to 1, for the 1PL, the common discrimination estimate was 5.371, which is quite large, indicating that the items are quite able to differentiate among those with higher levels of anxiety versus those with lower levels of anxiety. In addition, the item difficulty values were moderated in the 1PL model, with most of them being much nearer to 0, as compared with the Rasch model. The ICC and IIC for item 8 appear below.

Of particular note in Figures 9.7 and 9.8 is the much steeper ICC than was the case for the Rasch model and the migration of the IIC to a higher level on the *x*-axis. This means that item 8 provides maximum information for the 1PL at an anxiety level of approximately −0.5 (0.5 standard deviations below average) as compared with the −1.5 for the Rasch model.

Finally, we can fit the 2PL model to our data by altering the model command to the following:

```
MODEL:      anx BY anxq1* anxq2-anxq25;
```

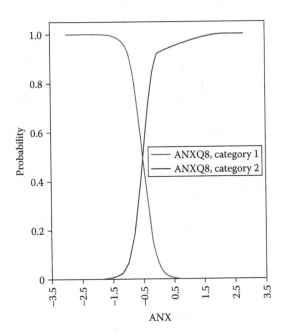

FIGURE 9.7
ICC for anxiety scale item 8, 1PL model.

The model fit indices for the 2PL model appear below.

```
MODEL FIT INFORMATION

Number of Free Parameters                        50

Loglikelihood

        H0 Value                         -2961.291
        H0 Scaling Correction Factor        0.9777
           for MLR

Information Criteria

        Akaike (AIC)                       6022.581
        Bayesian (BIC)                     6224.236
        Sample-Size Adjusted BIC           6065.572
           (n* = (n + 2) / 24)
```

Note here that no constraints are placed on the values of the discrimination parameters for the items. The information criteria are all lower for the 2PL model than for either the 1PL or the Rasch, indicating that it provides

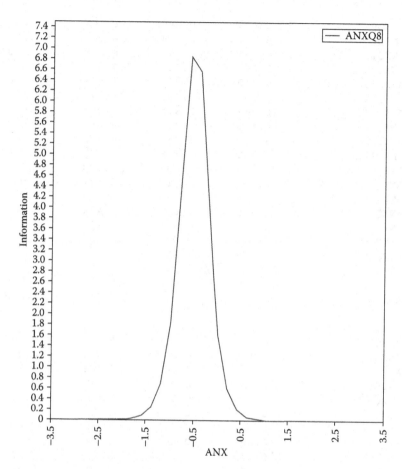

FIGURE 9.8
IIC for anxiety scale item 8, 1PL model.

optimal fit to the data, from among these three models. The item parameter estimates appear below.

```
IRT PARAMETERIZATION

Item Discriminations

ANX       BY
   ANXQ1              6.572       0.978       6.719       0.000
   ANXQ2             10.014       1.904       5.260       0.000
   ANXQ3              7.280       1.119       6.508       0.000
   ANXQ4              9.550       1.594       5.990       0.000
   ANXQ5              5.421       0.770       7.039       0.000
   ANXQ6              6.032       0.932       6.475       0.000
   ANXQ7              3.194       0.388       8.232       0.000
   ANXQ8              5.366       0.799       6.720       0.000
```

ANXQ9	5.833	0.845	6.905	0.000
ANXQ10	5.387	0.774	6.960	0.000
ANXQ11	6.897	1.073	6.428	0.000
ANXQ12	7.236	1.022	7.083	0.000
ANXQ13	8.849	1.325	6.679	0.000
ANXQ14	3.282	0.392	8.377	0.000
ANXQ15	10.049	1.915	5.248	0.000
ANXQ16	3.629	0.460	7.881	0.000
ANXQ17	8.368	1.335	6.270	0.000
ANXQ18	2.715	0.328	8.271	0.000
ANXQ19	4.910	0.682	7.198	0.000
ANXQ20	7.895	1.417	5.572	0.000
ANXQ21	7.992	1.366	5.852	0.000
ANXQ22	4.928	0.663	7.434	0.000
ANXQ23	4.995	0.694	7.198	0.000
ANXQ24	5.928	0.928	6.387	0.000
ANXQ25	9.908	1.850	5.357	0.000

Means
ANX	0.000	0.000	0.000	1.000

Item Difficulties
ANXQ1	-0.326	0.059	-5.541	0.000
ANXQ2	-0.190	0.059	-3.211	0.001
ANXQ3	-0.069	0.064	-1.075	0.282
ANXQ4	0.126	0.071	1.780	0.075
ANXQ5	0.126	0.075	1.677	0.094
ANXQ6	0.355	0.090	3.920	0.000
ANXQ7	-0.127	0.072	-1.748	0.081
ANXQ8	-0.500	0.061	-8.194	0.000
ANXQ9	-0.159	0.063	-2.536	0.011
ANXQ10	-0.060	0.067	-0.902	0.367
ANXQ11	0.023	0.068	0.338	0.736
ANXQ12	0.054	0.069	0.785	0.433
ANXQ13	0.167	0.073	2.273	0.023
ANXQ14	-0.038	0.075	-0.510	0.610
ANXQ15	0.226	0.076	2.985	0.003
ANXQ16	-0.463	0.067	-6.890	0.000
ANXQ17	-0.265	0.058	-4.533	0.000
ANXQ18	-0.597	0.076	-7.887	0.000
ANXQ19	-0.079	0.067	-1.186	0.236
ANXQ20	0.039	0.067	0.581	0.562
ANXQ21	0.180	0.075	2.404	0.016
ANXQ22	0.066	0.073	0.895	0.371
ANXQ23	0.114	0.075	1.512	0.131
ANXQ24	-0.053	0.066	-0.798	0.425
ANXQ25	0.207	0.075	2.764	0.006

Variances
ANX	1.000	0.000	0.000	1.000

Clearly, there are differences in the values of the discrimination parameters, all of which are quite large. This variation in the values of the a parameter estimates is certainly the reason that the 2PL model provides better fit to the data than either the Rasch or 1PL models. The 2PL ICC and IIC for item 8 appear in the following (Figures 9.9 and 9.10).

Interpretation of these graphs is carried out in much the same fashion as is the case for the other IRT models that we have examined in this chapter. Finally, given that endorsing an item on the scale due solely to chance (i.e., guessing) is not a reasonable option in this context, we would not fit a 3PL model to the data.

Mixture Models

In this chapter, we have focused on continuous latent variables, such as factors and the latent trait in IRT. However, it is also possible to model discrete latent variables consisting of a small number of categories in the population that are differentiated by some function involving a set of observed variables, using techniques collectively known as mixture models. The latent categories can be differentiated by the means of categorical or continuous variables

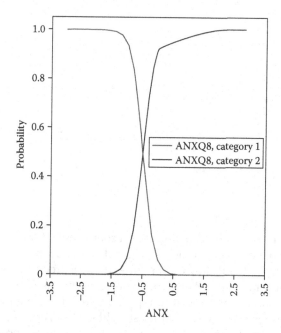

FIGURE 9.9
ICC for anxiety scale item 8, 2PL model.

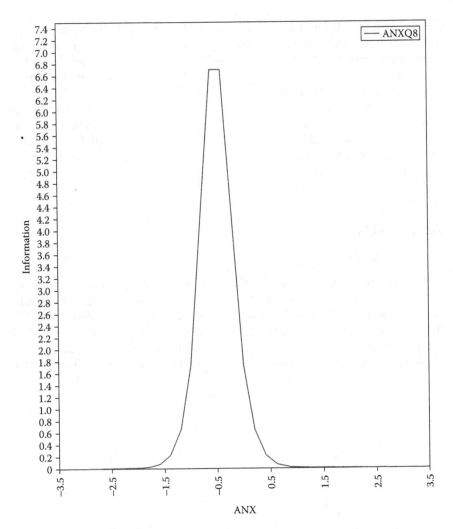

FIGURE 9.10
IIC for anxiety scale item 8, 2PL model.

or by different parameters (e.g., regression coefficients) for statistical models involving a set of observed variables. Regardless of how the classes are determined, it is assumed that they can be directly measured by using a function of the observed indicators. We should note here that as with the rest of the chapter, this section is not meant to represent an exhaustive presentation of mixture modeling in Mplus but rather should serve as a useful introduction on which the interested reader can build a broad base of knowledge about fitting these models.

Latent Class Models

Probably the most basic mixture model is latent class analysis (Lazarsfeld & Henry, 1968). It is typically conducted in an exploratory manner, much as is the case with EFA, where firmly grounded a priori hypotheses regarding the number of classes underlying the data may not be available (Collins & Lanza, 2010). Typically, the researcher will fit several models to the data, each being differentiated from the others by the number of latent classes that are modeled. The fit of these models is then compared by using a variety of indices, and the one yielding the best fit to the data is selected as being optimal. The nature of the latent classes can be ascertained through an examination of the means of the observed indicators for each class. Researchers typically use AIC, BIC, and aBIC to determine which model provides best fit to the data, along with model fit tests based on χ^2 and likelihood ratio statistics, including the Lo-Mendell-Rubin (LMR) (Lo, Mendell & Rubin, 2001; Nylund, Asparouhov, & Muthén, 2007) test and the bootstrap likelihood ratio test (BLRT) (McLachlan & Peel, 2000). The final determination as to the appropriateness of a model for a given set of data should also take into account the substantive coherence of the latent classes (Bauer & Curran, 2004), in much the same way that interpretability is a key criterion in determining the optimal number of factors in EFA.

The basic LCA model for observed indicator X and latent class variable Y is expressed as

$$\pi_{1,2,3\ldots,jt}^{X_1 X_2 X_3 \ldots X_j, Y} = \sum_{t}^{T} \pi_t^Y \Pi \pi_{jt}^{X_j|Y} \tag{9.19}$$

where:

π_t^Y is the probability that a randomly selected individual will be in class t of latent variable Y

$\pi_{jt}^{X_j|Y}$ is the probability that a member of latent class t will provide a particular response to observed indicator j

Model Equation 9.19 indicates that individual responses to the observed variables are conditionally independent of one another, given a particular latent class in Y. This model is characterized by two parameters: (1) the probability of a particular value for an observed variable conditional on latent class membership, and (2) the probability of an individual being in a specific latent class, t. Note that when the observed variables are continuous, $\pi_{jt}^{X_j|Y}$ are the means of these indicators, conditioned on the latent class membership.

Covariates of class membership can be incorporated into Model Equation 9.19, where the relationship between the latent class variable and the covariate is expressed as a logistic regression model. Here, z_k is a covariate that is believed to be associated with latent class membership.

$$\pi_{1,2,3\ldots,jt}^{X_1 X_2 X_3 \ldots X_j, Y} = \sum_{t}^{T} \left(\pi_t^Y | z_k \right) \Pi \pi_{jt}^{X_j|Y} \tag{9.20}$$

The inclusion of covariates into the model can yield a better understanding regarding the nature of the latent classes and can also improve the accuracy of the model in terms of correct recovery of the underlying latent classes in the population.

Fitting LCA by Using Mplus

Fitting the LCA model with or without covariates can be done very easily by using Mplus. As an example, let us consider a scenario in which a researcher has collected data from 555 high school sophomores on a variety of measures designed to assess various aspects of their social/political engagement, as well as their knowledge of civics content and their skills in reading and understanding political and government information. The researcher is interested in learning whether there are subgroups in the population based on score on these measures. The Mplus code that can be used to fit a 2-class LCA model appears below and is contained in the file `model_9_x_2class.inp`. We will fit several LCA models that differ in terms of the number of latent classes and then use statistical and conceptual considerations to determine the model that provides the optimal fit.

```
TITLE:   LCA for Civics data
DATA: FILE IS civics.csv;
VARIABLE:      NAMES ARE content
skills
civics
citizen
social
economy
security
trust
patriotism
women
immigration
school_A
political_A
climate
gender;

       USEVARIABLES ARE content
skills
civics
citizen
social
economy
security
trust
patriotism
```

```
women
immigration
school_A
political_A
climate;

CLASSES = c (2);

missing are .;

ANALYSIS:   TYPE = MIXTURE;
            lrtbootstrap=20;
OUTPUT:     TECH11 TECH14;
```

The first portion of this program is very similar to others that we have seen in previous examples. Note that missing data are indicated in the dataset by, and that we will not use the gender variable in the initial analysis. One of the new aspects of this program that we have not seen previously is the specification of the number of latent classes that we would like to fit, which we indicate in the line CLASSES = c (2);. In the ANALYSIS command, we specify that we are fitting a mixture model and that we would like 20 bootstrap samples for the bootstrap test of model fit. This is far fewer than we would actually fit in practice, but we set the number low in this case, so that the analysis will be completed in a timely fashion. The reader is encouraged to set this number much higher, perhaps to 1000, when he has ample time (and other tasks to occupy that time) to wait for the model to complete running. By specifying TECH11 and TECH14, we are requesting the LMR test of model fit (Lo, Mendell, & Rubin, 2001) and the bootstrap likelihood ratio test (BLRT) of model fit (McLachlan & Peel, 2000). Finally, we will request the ability to create some plots that might help us understand the nature of our latent classes once the optimal solution is decided upon.

In order to determine which model (i.e., number of latent classes) yields optimal fit to the data, we will examine from 2- to 5-latent-class solutions and rely on the BLRT and LMR, as well as the aBIC, per recommendations by Nylund, Asparouhov, and Muthén (2007). Below are the results for these statistics for the 2-class solution.

```
2 classes
          Sample-Size Adjusted BIC        39320.421
             (n* = (n + 2) / 24)
      LO-MENDELL-RUBIN ADJUSTED LRT TEST

             Value                             863.165
             P-Value                             0.0019
      PARAMETRIC BOOTSTRAPPED LIKELIHOOD RATIO TEST FOR 1 (H0)
VERSUS 2 CLASSES
```

```
HO Loglikelihood Value                              -20028.739
2 Times the Loglikelihood Difference                  872.271
Difference in the Number of Parameters                     15
Approximate P-Value                                    0.0000
Successful Bootstrap Draws                                 20
```

We will compare the aBIC values for each of the models that we fit, and the one with the smallest such value is taken to provide the best fit. This use of the aBIC is identical to that for other types of modeling applications that we have discussed in the book. The LMR is testing the null hypothesis that the fit of a 1-class model is equal to that of the 2-class model. Thus, when the *p*-value is less than α (e.g., 0.05), we conclude that the 2-class model provides better fit to the data than the 1-class model. The null hypothesis of the BLRT is the same as that of the LMR, so that *p*-value less than α indicates that the more complex model (the one with more latent classes) provides better fit to the data. For this example, both tests are statistically significant, indicating that the 2-class model fits the data better than the 1-class model. The results for the 3-class, 4-class, and 5-class models appear below.

```
3 classes
            Sample-Size Adjusted BIC       38844.762
               (n* = (n + 2) / 24)
      LO-MENDELL-RUBIN ADJUSTED LRT TEST

            Value                                    517.368
            P-Value                                   0.0029
      PARAMETRIC BOOTSTRAPPED LIKELIHOOD RATIO TEST FOR 2 (HO)
VERSUS 3 CLASSES

            HO Loglikelihood Value                -19592.603
            2 Times the Loglikelihood Difference     522.826
            Difference in the Number of Parameters        15
            Approximate P-Value                       0.0000
            Successful Bootstrap Draws                    20

4 classes
            Sample-Size Adjusted BIC       38481.455
               (n* = (n + 2) / 24)
      LO-MENDELL-RUBIN ADJUSTED LRT TEST

            Value                                    406.190
            P-Value                                   0.0060
      PARAMETRIC BOOTSTRAPPED LIKELIHOOD RATIO TEST FOR 3 (HO)
VERSUS 4 CLASSES

            HO Loglikelihood Value                -19331.190
            2 Times the Loglikelihood Difference     410.475
            Difference in the Number of Parameters        15
            Approximate P-Value                       0.0000
            Successful Bootstrap Draws                    20
```

```
5 classes

         Sample-Size Adjusted BIC          38267.666
           (n* = (n + 2) / 24)
      LO-MENDELL-RUBIN ADJUSTED LRT TEST

         Value                                258.232
         P-Value                               0.2789
      PARAMETRIC BOOTSTRAPPED LIKELIHOOD RATIO TEST FOR 4 (H0)
 VERSUS 5 CLASSES

         H0 Loglikelihood Value             -19125.953
         2 Times the Loglikelihood Difference   260.957
         Difference in the Number of Parameters    15
         Approximate P-Value                    0.0000
         Successful Bootstrap Draws               20
```

The results presented above present a mixed picture in terms of the optimal latent class solution. Based on aBIC and the BLRT, the 5-class model yields the best fit to the data, whereas the LMR would indicate that the 4-class model fits the data better than the 3-class model and the 5-class model does not yield better fit than the 4-class model. An additional consideration would be the conceptual quality of the results. In other words, how large are the resulting latent classes for each solution? Do the classes make sense from a conceptual perspective? Based on these considerations, the 4-class solution may provide the optimal solution. Not only do the latent classes appear to make substantive sense for the 4-class model, but also the smallest class for the 5-class solution contained only 22 individuals. Thus, it may be that this group may be too small for the researcher to fully trust as being generalizable to the broader population. For these reasons, we will focus on the results for the 4-class solution.

When interpreting the results from an LCA model, we will focus on several different pieces of information related to the parameters of interest, the proportion of individuals in each latent class, and the class-specific means of the observed indicators. Note that in this brief review, we will not discuss every portion of the output but rather focus on only those aspects that we feel are most relevant to understanding the results of the LCA. First, it is important to know the size of each class. The actual numbers of individuals assigned to each is presented in the following table.

```
FINAL CLASS COUNTS AND PROPORTIONS FOR THE LATENT CLASSES
BASED ON THEIR MOST LIKELY LATENT CLASS MEMBERSHIP

Class Counts and Proportions
```

```
Latent
Classes

   1                117            0.21081
   2                184            0.33153
   3                200            0.36036
   4                 54            0.09730
```

Clearly, there is some differentiation in terms of the class sizes, with the smallest representing just under 10% of the total sample and the largest being 36%.

A useful statistic for ascertaining the quality of the LCA solution is the mean probability of the most likely latent class, taken across individuals. Each individual in the sample is assigned a probability for membership in each of the latent classes and is placed in the class for which they have the largest probability. If the probability for the assigned class is very high (e.g., near 1), then we know that the model is confident about the class membership of that individual. Furthermore, if the mean assignment probability for each class is high, then we know that class separation is good and that the quality of the LCA solution is relatively higher than the case if the mean probability of class assignment is low. The mean class assignment probabilities are presented in the following table.

```
Average Latent Class Probabilities for Most Likely Latent
Class Membership (Row)
by Latent Class (Column)

            1          2          3          4

   1     0.958      0.000      0.042      0.000
   2     0.000      0.943      0.043      0.013
   3     0.030      0.030      0.940      0.000
   4     0.000      0.018      0.000      0.982
```

The mean probabilities for the assigned latent classes appear on the diagonal of the table. So, for example, the mean probability of being in latent class 1 for those who were assigned to latent class 1 was 0.958, indicating that these individuals were very likely assigned to the correct class. Indeed, for this example, all of the mean probabilities were 0.94 or higher, indicating a high degree of confidence in the assignment probabilities for this analysis. Entropy is a statistic that summarizes these latent class probabilities and is calculated as

$$E = 1 - \frac{\sum_{i=1}^{N}\sum_{c=1}^{C} -p_{ic}\ln p_{ic}}{N\ln p_c} \tag{9.21}$$

where:
  p_{ic} is the probability of subject i being in class c
  p_c is the probability of membership in class c
  N is the sample size

Values near 1 indicate better latent class separation or, put another way, a higher-quality latent class solution.

The results above would suggest that the latent classes are distinct in terms of the probabilities of class membership. It is also important for us to examine the nature of the classes in terms of the means of the variables used to create them. Mplus provides us with the class-specific means in the following output.


```
MODEL RESULTS
```

	Estimate	S.E.	Est./S.E.	Two-Tailed P-Value
Latent Class 1				
Means				
CONTENT	76.778	0.982	78.208	0.000
SKILLS	85.027	1.713	49.634	0.000
CIVICS	78.504	1.128	69.625	0.000
CITIZEN	10.370	0.254	40.876	0.000
SOCIAL	9.550	0.256	37.244	0.000
ECONOMY	9.031	0.196	46.186	0.000
SECURITY	8.713	0.228	38.250	0.000
TRUST	10.267	0.247	41.545	0.000
PATRIOTISM	9.489	0.253	37.468	0.000
WOMEN	8.923	0.216	41.375	0.000
IMMIGRATIO	9.366	0.237	39.529	0.000
SCHOOL_A	9.449	0.235	40.123	0.000
POLITICAL_A	10.695	0.264	40.519	0.000
CLIMATE	9.790	0.224	43.776	0.000
Variances				
CONTENT	63.685	5.171	12.315	0.000
SKILLS	162.659	12.080	13.465	0.000
CIVICS	54.133	3.685	14.690	0.000
CITIZEN	5.261	0.488	10.787	0.000
SOCIAL	5.134	0.329	15.599	0.000
ECONOMY	3.700	0.314	11.768	0.000
SECURITY	3.349	0.212	15.768	0.000
TRUST	4.034	0.399	10.108	0.000
PATRIOTISM	4.735	0.264	17.946	0.000
WOMEN	4.135	0.234	17.697	0.000

IMMIGRATIO	4.594	0.305	15.044	0.000
SCHOOL_A	4.347	0.265	16.400	0.000
POLITICAL_A	3.934	0.255	15.455	0.000
CLIMATE	4.905	0.404	12.145	0.000

Latent Class 2

Means

CONTENT	114.261	1.323	86.360	0.000
SKILLS	129.041	1.195	107.982	0.000
CIVICS	119.920	1.205	99.489	0.000
CITIZEN	10.255	0.208	49.309	0.000
SOCIAL	10.230	0.195	52.550	0.000
ECONOMY	8.930	0.164	54.451	0.000
SECURITY	9.989	0.173	57.719	0.000
TRUST	10.542	0.140	75.376	0.000
PATRIOTISM	9.705	0.185	52.385	0.000
WOMEN	10.753	0.193	55.729	0.000
IMMIGRATIO	10.221	0.177	57.855	0.000
SCHOOL_A	10.281	0.184	55.982	0.000
POLITICAL_A	10.739	0.142	75.877	0.000
CLIMATE	10.607	0.205	51.665	0.000

Variances

CONTENT	63.685	5.171	12.315	0.000
SKILLS	162.659	12.080	13.465	0.000
CIVICS	54.133	3.685	14.690	0.000
CITIZEN	5.261	0.488	10.787	0.000
SOCIAL	5.134	0.329	15.599	0.000
ECONOMY	3.700	0.314	11.768	0.000
SECURITY	3.349	0.212	15.768	0.000
TRUST	4.034	0.399	10.108	0.000
PATRIOTISM	4.735	0.264	17.946	0.000
WOMEN	4.135	0.234	17.697	0.000
IMMIGRATIO	4.594	0.305	15.044	0.000
SCHOOL_A	4.347	0.265	16.400	0.000
POLITICAL_A	3.934	0.255	15.455	0.000
CLIMATE	4.905	0.404	12.145	0.000

Latent Class 3

Means

CONTENT	93.627	1.139	82.216	0.000
SKILLS	111.172	1.556	71.426	0.000
CIVICS	98.697	1.122	87.980	0.000
CITIZEN	10.320	0.201	51.254	0.000
SOCIAL	10.504	0.175	60.141	0.000
ECONOMY	9.445	0.170	55.530	0.000
SECURITY	10.044	0.161	62.404	0.000
TRUST	10.308	0.181	56.879	0.000

PATRIOTISM	10.283	0.182	56.452	0.000
WOMEN	10.470	0.165	63.306	0.000
IMMIGRATIO	10.655	0.171	62.151	0.000
SCHOOL_A	10.285	0.178	57.649	0.000
POLITICAL_A	10.242	0.166	61.833	0.000
CLIMATE	10.656	0.190	55.938	0.000

Variances

CONTENT	63.685	5.171	12.315	0.000
SKILLS	162.659	12.080	13.465	0.000
CIVICS	54.133	3.685	14.690	0.000
CITIZEN	5.261	0.488	10.787	0.000
SOCIAL	5.134	0.329	15.599	0.000
ECONOMY	3.700	0.314	11.768	0.000
SECURITY	3.349	0.212	15.768	0.000
TRUST	4.034	0.399	10.108	0.000
PATRIOTISM	4.735	0.264	17.946	0.000
WOMEN	4.135	0.234	17.697	0.000
IMMIGRATIO	4.594	0.305	15.044	0.000
SCHOOL_A	4.347	0.265	16.400	0.000
POLITICAL_A	3.934	0.255	15.455	0.000
CLIMATE	4.905	0.404	12.145	0.000

Latent Class 4

Means

CONTENT	143.154	1.986	72.095	0.000
SKILLS	136.996	1.590	86.178	0.000
CIVICS	149.975	2.781	53.933	0.000
CITIZEN	10.595	0.340	31.176	0.000
SOCIAL	9.956	0.372	26.746	0.000
ECONOMY	8.828	0.316	27.979	0.000
SECURITY	10.470	0.303	34.600	0.000
TRUST	10.871	0.190	57.327	0.000
PATRIOTISM	10.030	0.347	28.939	0.000
WOMEN	11.226	0.309	36.275	0.000
IMMIGRATIO	10.591	0.283	37.422	0.000
SCHOOL_A	10.569	0.321	32.939	0.000
POLITICAL_A	11.393	0.204	55.960	0.000
CLIMATE	10.627	0.349	30.437	0.000

Variances

CONTENT	63.685	5.171	12.315	0.000
SKILLS	162.659	12.080	13.465	0.000
CIVICS	54.133	3.685	14.690	0.000
CITIZEN	5.261	0.488	10.787	0.000
SOCIAL	5.134	0.329	15.599	0.000
ECONOMY	3.700	0.314	11.768	0.000
SECURITY	3.349	0.212	15.768	0.000
TRUST	4.034	0.399	10.108	0.000

PATRIOTISM	4.735	0.264	17.946	0.000
WOMEN	4.135	0.234	17.697	0.000
IMMIGRATIO	4.594	0.305	15.044	0.000
SCHOOL_A	4.347	0.265	16.400	0.000
POLITICAL_A	3.934	0.255	15.455	0.000
CLIMATE	4.905	0.404	12.145	0.000

When examining these tables, we are looking for patterns of the means of the observed variables that differentiate the groups from one another. For example, individuals in class 4 had the highest means on the tests of content, skills, and civics, followed by members of class 2. Those in latent class 1 had the lowest mean scores for these measures. Thus, we can conclude that individuals in latent class 4 were the most knowledgeable and those in class 1 had the least knowledge about public affairs. In addition, members of class 1 had relatively higher scores for trust in institutions and political activism, when compared with the other measures. Latent class 2 was characterized by having somewhat lower scores for the government's role in the economy, security, and patriotism. Class 3 had relatively similar (and high) means for all of the measures, except for the government's role in the economy. Finally, latent class 4 had relatively high means for all of the measures, except for the government's role in the economy, and also had the largest means of all classes for both women's rights and engagement in political activities. Taken together, it would appear that latent class 4 tends to be very knowledgeable about civics issues and highly engaged in political activities, with particular interest in women's rights. Class 1 is relatively less informed and less engaged than the other groups, whereas class 2 is relatively well informed and somewhat less interested in the government taking a role in the economy than the members of the other classes. Finally, class 3 is not as well informed as classes 2 and 4 but is more informed than class 1.

As noted earlier, it is possible to include a covariate predicting latent class membership into the LCA model itself. For this example, we will extend the model by including gender (1 = female, 0 = male) as such a covariate, in order to determine whether gender is related to class membership. In order to do so, we will use model_9_x_4class_covariate.inp, which appears below.

```
TITLE: LCA for Civics data with covariate
DATA: FILE IS civics.csv;
VARIABLE:    NAMES ARE content
skills
civics
citizen
social
economy
security
trust
```

```
patriotism
women
immigration
school_A
political_A
climate
gender;

        USEVARIABLES ARE content
skills
civics
citizen
social
economy
security
trust
patriotism
women
immigration
school_A
political_A
climate
gender;

CLASSES = c (4);

missing are .;

ANALYSIS:     TYPE = MIXTURE;
!             lrtbootstrap=20;

model:        %overall%
              c on gender;

OUTPUT:       TECH11 TECH14;
```

The primary difference between this program and the previous one is that here, we include gender in the USEVARIABLES command and we add the MODEL statement, linking class membership to gender. This additional command instructs Mplus to fit a multinomial logistic regression model in which latent class membership is the dependent variable and gender is the independent variable. By default, the last latent class (4) serves as the reference class, against which the others are compared. Before examining the logistic regression results, we may first want to compare the fit of this model to the one without the covariate, using the aBIC value. For the 4-class LCA model with the covariate, aBIC is 38477.053, whereas for the 4-class model without the covariate, the aBIC value is 38481.455. Thus, inclusion of the covariate and associated logistic regression model yields a somewhat better fit to the data. The results for the logistic regression involving gender appear below.

```
Categorical Latent Variables

C#1            ON
    GENDER              -0.840        0.270       -3.108        0.002


C#2            ON
    GENDER              -0.729        0.338       -2.155        0.031


C#3            ON
    GENDER              -0.220        0.247       -0.891        0.373

Intercepts
    C#1                 -0.122        0.192       -0.634        0.526
    C#2                 -0.928        0.230       -4.034        0.000
    C#3                  0.018        0.191        0.095        0.925



LOGISTIC REGRESSION ODDS RATIO RESULTS

Categorical Latent Variables

C#1        ON
    GENDER              0.432


C#2        ON
    GENDER              0.482


C#3        ON
    GENDER              0.803
```

One important point to note here is that inclusion of a covariate and the associated logistic regression into the model will likely alter the LCA results to some extent. Indeed, such is the case here. Latent class 1 remains the lower-achieving, less-engaged group, but latent class 2 is now the highest-achieving (formerly class 4), highly engaged group; latent class 3 is the next highest-achieving group (formerly class 3); and latent class 4 (formerly class 2) is the next lowest-achieving group but one that was relatively engaged in politics and social activism. From the logistic regression results, we see that gender is related to class membership, with respect to both class 1 versus class 4 and class 2 versus class 4. Recall that females are coded as 1 and males are coded as 0. Therefore, the statistically significant negative coefficients for gender on classes 1 and 2 mean that females were less likely to belong to either class 1 or class 2 than they were to class 4. Mplus also provides us with the odds ratios for these genders by these classes, and we can see that females are less than half as likely as males to belong to either class 1 or 2 versus class 4.

Summary

The purpose of this chapter was to introduce the basics for using Mplus to fit a variety of latent variable models, including factor analysis, structural equation modeling, growth curve modeling, item response theory, and latent class analysis. Clearly, this is only an introduction, as each of these topics could easily warrant a book of its own. However, we are hopeful that the reader who was unfamiliar with how to fit such models by using Mplus now is more confident in doing so. In addition, this chapter was designed to serve as a bridge to the next chapter, which focuses on the application of Mplus to fit multilevel latent variable models. Most of the core ideas that were introduced here will be used in the multilevel context, with some expansions to accommodate the additional level(s) of data.

10

Multilevel Latent Variable Models in Mplus

As with the other data analysis paradigms that we have described in this book (e.g., linear models), latent variable models can also be modeled in the multilevel context. For example, researchers may collect quality of life data from cancer patients visiting several different clinics, with an interest in modeling the factor structure of these scales. Given that the data were collected from clustered groups of individuals, the standard approaches to modeling them (e.g., exploratory factor analysis [EFA] and confirmatory factor analysis [CFA]), which assume independence of the observations, must be adapted to account for the additional structure in the data. Similarly, researchers may wish to identify a small number of latent classes in the population, based on the quality of life scale responses. Again, using the standard method for fitting latent class analysis (LCA) models that was described in Chapter 9 would be inappropriate in this case, because the observations are not independent of one another. However, multilevel LCA (MLCA) modeling of such data is an option and can be effected relatively easily by using Mplus, as we will see below. The purpose of this chapter is to describe the fitting of multilevel versions of the models that we first described in Chapter 9. Although we endeavor to present a wide array of such models, we also recognize that in one chapter, we will only be able to touch on the main points. The interested reader is encouraged to investigate the more advanced aspects of multilevel latent variable modeling through the references that appear below.

Multilevel Factor Analysis

In Chapter 9, we described the factor analysis (FA) model and methods for fitting it by using Mplus. Recall that FA is useful in situations where researchers are interested in understanding the latent structure underlying observed indicator variables, such as items on a scale or subscales from an instrument; for example, FA was described as being conducted in either of the two ways, exploratory (EFA) or confirmatory (CFA). With EFA, the researcher does not place any constraints on the model in terms of which observed indicators are associated with which latent variables, though she may specify the number of factors (or a range for the number of factors) that should be fit. The primary goal when conducting EFA is to ascertain the optimal (and hopefully correct with respect to the population) number and nature of the factors. Thus, the focus is on selecting the number of factors to retain and on interpreting the factors, based on which indicators are primarily associated with (load on) each. In contrast, CFA modeling begins with the researcher specifying the factor model structure by using a set of constraints, typically with respect to which variables are associated with which factors, through the loadings. Both multilevel EFA and CFA models can be fit by using Mplus.

The single-level factor model was introduced in Equation 9.5 as

$$y_i = \tau + \Lambda\eta + \varepsilon_i \tag{10.1}$$

where:

y_i is vector of observed indicator variables for individual i
τ is factor intercept
η is latent variable(s)
Λ is factor loadings linking observed indicators with factors
ε_i is unique variances for the indicators for individual i

A primary assumption of this model is that the ε_i are independent of one another. However, when individuals are sampled in a clustered fashion (e.g., patients at the same clinic), this assumption is no longer valid. As a result of violating this assumption, estimates of the parameters in Equation 10.1 that are based on the covariance matrix naively not accounting for the additional structure will have the same types of problems associated with them that we described in earlier chapters with reference to naïve linear models fit to multilevel data (Muthén, 1989). Put another way, the covariance matrix for the observed indicator variables in the multilevel context is a function of both within- and between-cluster sources of variation, which can be expressed as

$$\Sigma = \Sigma_W + \Sigma_B \tag{10.2}$$

where:

Σ_W is within-groups covariance matrix
Σ_B is between-groups covariance matrix

The covariance matrices in Equation 10.2 can be expressed directly in terms of factor model parameters. The overall covariance matrix can be expressed in terms of the overall model parameters as Equation 10.3, which is simply a restatement of Equation 9.2.

$$\Sigma = \Lambda\Psi\Lambda' + \Theta \tag{10.3}$$

where:

Σ is model-predicted correlation matrix of the indicators
Ψ is correlation matrix for the factors
Θ is diagonal matrix of unique error variances

Each component of Equation 10.2 can also be described using the appropriate (within or between) factor model parameters. For example, Σ_W is a function of within-cluster loadings, variances, and covariances.

$$\Sigma_W = \Lambda_W \Psi_W \Lambda'_W + \Theta_W \tag{10.4}$$

The terms in Equation 10.4 are identical to those in Equation 10.3, except that they are at the within-cluster level. Similarly, we can express the between-cluster covariance matrix in terms of between-cluster level factor model parameters.

$$\Sigma_B = \Lambda_B \Psi_B \Lambda'_B + \Theta_B \tag{10.5}$$

Together, Equations 10.4 and 10.5 imply that Equation 10.1 can be extended to the multilevel framework by combining the within and between portions of the covariance matrix as in Equation 10.6.

$$y_{ic} = \tau_B + \Lambda_B \eta_{Bc} + \Lambda_{Wic} \eta_{Wic} + \varepsilon_B + \varepsilon_{Wic} \tag{10.6}$$

The terms in Equation 10.6 are as defined previously, with c indicating cluster membership. We can see that the level 1 intercepts (τ_W) are set to 0, because the value of y for individual i in cluster c is due to a unique deviation from the overall cluster c mean for that indicator. In addition, the multilevel FA model also has associated with it factor loadings at both levels of the data. In other words, there will be factor loadings both at the cluster level (Λ_B) and at the individual level (Λ_{Wic}). Likewise, factor scores are also estimated at both levels, as are the error terms.

One additional aspect of multilevel FA modeling that is implied by Equations 10.4 and 10.5 is the decomposition of the total factor variance (Ψ) into the portion due to variation between clusters (Ψ_B) and the portion due to variation within clusters (Ψ_W). The intraclass correlation (ICC) for the indicators can be calculated using these variances, as

$$\frac{\Psi_B}{\Psi_B + \Psi_W} \tag{10.7}$$

The interpretation of this value, as the proportion of variation in the observed variables that is a function of cluster membership, is the same as what we have discussed in prior chapters. In order to make the factor variances at levels 1 and 2 comparable with one another, Mehta and Neale (2005) note that the factor loadings across the levels must be invariant with one another. This means that the researcher needs to constrain the loading at level 1 for a given indicator to be equivalent to the loading at level 2 for that same indicator. We will demonstrate how to do this below, using Mplus.

Fitting a Multilevel EFA Model by Using Mplus

In order to demonstrate the fitting of a multilevel EFA model by using Mplus, let us consider an example involving a quality of life scale that is given to 2000 cancer patients across 60 treatment centers. The instrument consists of 12 subscales that can be grouped into three latent variables, with subscales 1–4 associated with the first factor, subscales 5–8 associated with the second factor, and subscales 9–12 associated with the third factor. The data are contained in the file mlm_sem.dat, and the Mplus program is in the file model_10.1.inp, which appears below.

```
TITLE: Model 10.1  two level EFA

DATA: FILE IS mlm_sem.dat;

VARIABLE: NAMES ARE y1-y12 center;
USEVARIABLES = y1-y12;
CLUSTER = center;

ANALYSIS: TYPE = TWOLEVEL EFA 1 3 UW 1 3 UB;
estimator=ml;

output: residual;
```

Much of the program is structured as others that we have examined in this book. Of particular interest in the case of multilevel EFA is that we must indicate the factor structure that we would like to fit at both levels of the analysis, which we do in the ANALYSIS: line. In this case, we request two-level analysis for both levels 1 and 2, by using the command structure

```
EFA 1 3 UW 1 3 UB;
```

Although we did not do so here, we could have requested a different number of factors at each level. Finally, because we did not specify the rotation, Mplus will use Geomin.

The tools available to help us determine the number of factors are somewhat limited in the multilevel case. For example, with Mplus, we neither have access to parallel analysis nor to the scree plot. We can, however, request the residual correlation matrix and can refer to the Chi-square goodness of fit test, as well as to the eigenvalues. One final point to note here is that Mplus will fit all combinations of the level 1 and level 2 factor structures (e.g., 1 between, 1 within; 2 between, 1 within; and 1 between, 2 within). Thus, a great deal of output can be created when fitting this model with a large number of factors. One tool that we might use in sorting through all of these results is the information indices. Table 10.1 includes the Akaike information criterion (AIC), Bayesian information criterion (BIC), sample size adjusted BIC (aBIC) values for each combination of level 1 and level 2 factors.

TABLE 10.1

Information Indices for Multilevel EFA

Model	AIC	BIC	ABIC
1 W 1 B	76632.271	76968.325	76777.702
2 W 1 B	75794.878	76192.542	75966.972
3 W 1 B	75619.681	76073.354	75816.013
1 W 2 B	76300.471	76698.135	76472.564
2 W 2 B	75516.518	75975.792	75715.273
3 W 2 B	75342.015	75857.298	75565.009

The multilevel EFA models with three between factors were not able to converge by using maximum likelihood (ML) estimation, which is why it does not appear in this table. We could pursue this issue further by changing the number of iterations and/or the starting values that we use in fitting the model. We could also try using a different estimator, such as a robust ML (e.g., MLMV in model_10.1b.inp). However, in exploring these alternatives, we were not able to achieve convergence for the models with three level 2 factors. Therefore, for pedagogical purposes, we will focus on the results that did achieve convergence, corresponding to the models in Table 10.1. Results from each of the information indices suggested that the model with three level 1 and two level 2 factors (3W2B) yielded the best fit to the data. The fit information for this model appears below.

```
EXPLORATORY FACTOR ANALYSIS WITH 3 WITHIN FACTOR(S) AND 2
BETWEEN FACTOR(S):


MODEL FIT INFORMATION

Number of Free Parameters                      92

Loglikelihood

        H0 Value                         -37579.007
        H1 Value                         -37527.170

Information Criteria

        Akaike (AIC)                      75342.015
        Bayesian (BIC)                    75857.298
        Sample-Size Adjusted BIC          75565.009
          (n* = (n + 2) / 24)
```

```
Chi-Square Test of Model Fit

        Value                              103.674
        Degrees of Freedom                      76
        P-Value                             0.0192
```

The Chi-square model fit test is statistically significant, indicating that the model predicted covariance matrix is not equal to the observed covariance matrix. However, the residual correlations, which appear below, would suggest that the fit of the model is good, given that none of the residuals exceeded 0.05.

```
            Residuals for Correlations
                Y1          Y2          Y3          Y4          Y5

            _____    _____    _____    _____    _____

    Y1        0.000
    Y2        0.002       0.000
    Y3       -0.011      -0.010       0.000
    Y4       -0.008      -0.010       0.000       0.000
    Y5       -0.002      -0.010      -0.002       0.007       0.000
    Y6       -0.008       0.016      -0.013      -0.003       0.002
    Y7       -0.003      -0.017       0.015      -0.003      -0.001
    Y8        0.008       0.006      -0.008      -0.021      -0.003
    Y9        0.017       0.003       0.024       0.013       0.012
    Y10      -0.003       0.034       0.001       0.015       0.002
    Y11       0.009       0.021      -0.001       0.017       0.002
    Y12       0.004       0.015       0.007       0.019      -0.008


            Residuals for Correlations
                Y6          Y7          Y8          Y9          Y10

            _____    _____    _____    _____    _____

    Y6        0.000
    Y7       -0.005       0.000
    Y8        0.001       0.006       0.000
    Y9       -0.005       0.004       0.001       0.000
    Y10       0.002       0.021       0.005      -0.007       0.000
    Y11       0.038      -0.018      -0.004      -0.002      -0.022
    Y12       0.010       0.010       0.021      -0.007      -0.007


            Residuals for Correlations
                Y11         Y12

            _____    _____

    Y11       0.000
    Y12      -0.018       0.000
```

The factor loadings for level 1 appear below.

WITHIN LEVEL RESULTS

GEOMIN ROTATED LOADINGS (* significant at 5% level)

	1	2	3
Y1	0.719*	0.010	0.014
Y2	0.734*	-0.038	0.026
Y3	0.672*	0.063	-0.054
Y4	0.729*	-0.005	-0.003
Y5	-0.003	0.719*	-0.032
Y6	-0.003	0.560*	0.012
Y7	0.003	0.521*	0.015
Y8	0.039	0.496*	0.036
Y9	0.006	0.013	0.689*
Y10	-0.041	0.002	0.374*
Y11	-0.049	0.086	0.255*
Y12	0.032	-0.015	0.472*

GEOMIN FACTOR CORRELATIONS (* significant at 5% level)

	1	2	3
1	1.000		
2	0.607*	1.000	
3	0.188*	0.495*	1.000

These results show that at level 1, the subscales load together, as theory would suggest that they should. In addition, there were no cross-loadings present. The correlations among the factors were all positive and statistically significant, with the largest relationship occurring between factor 1 and factor 2. The level 2 factor loadings are shown next.

BETWEEN LEVEL RESULTS

GEOMIN ROTATED LOADINGS (* significant at 5% level)

	1	2
Y1	1.020*	-0.042
Y2	1.000*	-0.003
Y3	1.063*	-0.148*
Y4	0.998*	0.003
Y5	0.015	0.992*
Y6	-0.004	1.001*

Y7	0.078	0.931*
Y8	-0.043	1.020*
Y9	0.789*	0.312*
Y10	0.831*	0.244*
Y11	0.836*	0.045
Y12	0.786*	0.303*

GEOMIN FACTOR CORRELATIONS (* significant at 5% level)

	1	2
1	1.000	
2	0.487*	1.000

The results at level 2 were somewhat less clear than those at level 1. In particular, while subscales 1–4 loaded together on factor 1 and subscales 5–8 loaded on factor 2, subscales 9–12 loaded primarily on factor 1 but also cross-loaded on factor 2 in some instances. Thus, at level 2, the solution is not as clear-cut as was true for level 1. The difficulty in fitting the model with three factors at level 2 may be the result of the smaller level 2 sample size when compared with that of level 1. Thus, the algorithm is attempting to fit the model with relatively few (60) data points, as compared with the fitting of the level 1 model with 2000 data points.

Fitting a Multilevel CFA Model by Using Mplus

Having examined the multilevel EFA model, let us turn our attention to fitting a multilevel CFA model by using Mplus. To do so, we will use the mlm_sem. dat file that was the focus of the EFA example. Recall that subscales 1–4 were hypothesized to belong to the first factor, subscales 5–8 to the second factor, and subscales 9–12 to the third factor. Interpreting fit of a multilevel latent variable model is more complex than is the case for the single-level model, because we must consider both levels. However, with the exception of Standardized Root Mean Square Residual (SRMR), the fit statistics that we typically rely upon combine information about model fit at both levels of the data. Because there are typically many more data points at level 1 in this example, these indices primarily reflect the fit of the model at level 1 (Ryu and West, 2009). Stapleton (2013) provides an excellent step-by-step example of how a researcher can assess fit of the model at each level, which we will be using here.

Based on Stapleton's (2013) guidelines, we first must obtain the Chi-square model fit statistics for the baseline models at levels 1 and 2, respectively. In order to obtain the baseline value for the level 1 (within) portion of the model, we must set the covariances of the observed indicators to 0 at level 1 and allow them to be estimated freely at level 2. This is done in the program `model_10.2_within_baseline.inp`, which appears below.

```
TITLE: Model 10.2  two level CFA level 1 baseline

DATA: FILE IS mlm_sem.dat;

VARIABLE: NAMES ARE y1-y12 center;
USEVARIABLES = y1-y12;
CLUSTER = center;

ANALYSIS: TYPE = TWOLEVEL;
estimator=ml;

MODEL:

        %WITHIN%
        y1 with y2@0 y3@0 y4@0 y5@0 y6@0 y7@0 y8@0
            y9@0 y10@0 y11@0 y12@0;
        y2 with y3@0 y4@0 y5@0 y6@0 y7@0 y8@0
            y9@0 y10@0 y11@0 y12@0;
        y3 with y4@0 y5@0 y6@0 y7@0 y8@0
            y9@0 y10@0 y11@0 y12@0;
        y4 with y5@0 y6@0 y7@0 y8@0
            y9@0 y10@0 y11@0 y12@0;
        y5 with y6@0 y7@0 y8@0
            y9@0 y10@0 y11@0 y12@0;
        y6 with y7@0 y8@0
            y9@0 y10@0 y11@0 y12@0;
        y7 with y8@0
            y9@0 y10@0 y11@0 y12@0;
        y8 with y9@0 y10@0 y11@0 y12@0;
        y9 with y10@0 y11@0 y12@0;
        y10 with y11@0 y12@0;
        y11 with y12@0;

        %BETWEEN%
        y1 with y2-y12;
        y2 with y3-y12;
        y3 with y4-y12;
        y4 with y5-y12;
        y5 with y6-y12;
        y6 with y7-y12;
        y7 with y8-y12;
        y8 with y9-y12;
        y9 with y10-y12;
        y10 with y11-y12;
        y11 with y12;

output: standardized;
```

The Chi-square fit statistic appears below.

```
Chi-Square Test of Model Fit

        Value                             4637.366
        Degrees of Freedom                      66
        P-Value                             0.0000
```

Similarly, we need to obtain the Chi-square fit statistic for the baseline level 2 model, which is done by using the program contained in model_10.2_ between_baseline.inp.

```
TITLE: Model 10.2  two level CFA

DATA: FILE IS mlm_sem.dat;

VARIABLE: NAMES ARE y1-y12 center;
USEVARIABLES = y1-y12;
CLUSTER = center;

ANALYSIS: TYPE = TWOLEVEL;
estimator=ml;

MODEL:

    %WITHIN%
    y1 with y2-y12;
    y2 with y3-y12;
    y3 with y4-y12;
    y4 with y5-y12;
    y5 with y6-y12;
    y6 with y7-y12;
    y7 with y8-y12;
    y8 with y9-y12;
    y9 with y10-y12;
    y10 with y11-y12;
    y11 with y12;

    %BETWEEN%
    y1 with y2@0 y3@0 y4@0 y5@0 y6@0 y7@0 y8@0
        y9@0 y10@0 y11@0 y12@0;
    y2 with y3@0 y4@0 y5@0 y6@0 y7@0 y8@0
        y9@0 y10@0 y11@0 y12@0;
    y3 with y4@0 y5@0 y6@0 y7@0 y8@0
        y9@0 y10@0 y11@0 y12@0;
    y4 with y5@0 y6@0 y7@0 y8@0
        y9@0 y10@0 y11@0 y12@0;
    y5 with y6@0 y7@0 y8@0
        y9@0 y10@0 y11@0 y12@0;
```

```
   y6 with y7@0 y8@0
      y9@0 y10@0 y11@0 y12@0;
   y7 with y8@0
      y9@0 y10@0 y11@0 y12@0;
   y8 with y9@0 y10@0 y11@0 y12@0;
   y9 with y10@0 y11@0 y12@0;
   y10 with y11@0 y12@0;
   y11 with y12@0;

output: standardized;
```

The resulting Chi-square fit statistic is:

```
Chi-Square Test of Model Fit
         Value                        874.140
         Degrees of Freedom               66
         P-Value                      0.0000
```

Continuing with Stapleton's steps for obtaining fit statistics at each level, we now must fit a model that is saturated at level 2 (thereby yielding perfect fit at that level), with the posited model appearing at level 1. This model is fit, using the program: model_10.2_between_saturated.inp.

```
TITLE: Model 10.2  two level CFA level 2 saturated

DATA: FILE IS mlm_sem.dat;

VARIABLE: NAMES ARE y1-y12 center;
USEVARIABLES = y1-y12;
CLUSTER = center;

ANALYSIS: TYPE = TWOLEVEL;
estimator=ml;

MODEL:

  %WITHIN%
  f1w BY y1-y4;
  f2w by y5-y8;
  f3w by y9-y12;

  %BETWEEN%
  y1 with y2-y12;
  y2 with y3-y12;
  y3 with y4-y12;
  y4 with y5-y12;
  y5 with y6-y12;
  y6 with y7-y12;
  y7 with y8-y12;
```

```
     y8 with y9-y12;
     y9 with y10-y12;
     y10 with y11-y12;
     y11 with y12;

output: standardized;
```

The resulting Chi-square fit statistic appears below.

```
Chi-Square Test of Model Fit

        Value                            45.389
        Degrees of Freedom                  51
        P-Value                         0.6952
```

Using an equation described by Ryu and West (2009), we can calculate the comparative fit index (CFI) for the level 1 portion of the model as

$$\text{CFI}_{\text{level 1}} = 1 - \frac{\max\left[\chi^2_{\text{level 2 saturated}} - df_{\text{level 2 saturated}}, 0\right]}{\max\left[\chi^2_{\text{level 1 baseline}} - df_{\text{level 1 baseline}}, 0\right]} \tag{10.8}$$

For this problem, the level 1 CFI would be

$$\text{CFI}_{\text{level 1}} = 1 - \frac{\max\left[45.389 - 51, 0\right]}{\max\left[4637.66 - 66, 0\right]} = 1 - \frac{0}{4571.66} = 1$$

Thus, we would conclude that the fit of the model at level 1 is extremely good.

Likewise, we can calculate the level 2 CFI value in a similar fashion, by obtaining the Chi-square goodness of fit statistic for the level 1 saturated model.

```
TITLE: Model 10.2  two level CFA level 2 saturated

DATA: FILE IS mlm_sem.dat;

VARIABLE: NAMES ARE y1-y12 center;
USEVARIABLES = y1-y12;
CLUSTER = center;

ANALYSIS: TYPE = TWOLEVEL;
estimator=ml;

MODEL:

        %WITHIN%
        y1 with y2-y12;
```

```
y2 with y3-y12;
y3 with y4-y12;
y4 with y5-y12;
y5 with y6-y12;
y6 with y7-y12;
y7 with y8-y12;
y8 with y9-y12;
y9 with y10-y12;
y10 with y11-y12;
y11 with y12;

%BETWEEN%
f1b BY y1-y4;
f2b by y5-y8;
f3b by y9-y12;
```

output: standardized;

This model yields the following Chi-square fit statistic.

Chi-Square Test of Model Fit

Value	35.497
Degrees of Freedom	51
P-Value	0.9513

We can then calculate the CFI for level 2 as

$$CFI_{level\ 2} = 1 - \frac{max\left[\chi^2_{level\ 1\ saturated} - df_{leve\ 1\ saturated}, 0\right]}{max\left[\chi^2_{level\ 2\ baseline} - df_{level\ 2\ baseline}, 0\right]}$$ (10.9)

$$CFI_{level\ 1} = 1 - \frac{max[35.497 - 51, 0]}{max[874.14 - 66, 0]} = 1 - \frac{0}{808.14} = 1$$

The fit at level 2 is extremely good, as it was at level 1.

The Mplus program to fit the full model at both levels is in the file model_10.2.inp and appears below.

```
TITLE: Model 10.2  two level CFA

DATA: FILE IS mlm_sem.dat;

VARIABLE: NAMES ARE y1-y12 center;
USEVARIABLES = y1-y12;
CLUSTER = center;

ANALYSIS: TYPE = TWOLEVEL;
estimator=ml;
```

```
MODEL:

    %WITHIN%
    f1w BY y1-y4;
    f2w by y5-y8;
    f3w by y9-y12;

    %BETWEEN%
    f1b BY y1-y4;
    f2b by y5-y8;
    f3b by y9-y12;

output: standardized;
```

Much of this program is very similar to the more constrained versions that we just examined. The subscales are labeled y1–y12, and the treatment center is noted by the variable center. By default, estimation will be done using ML, which is what we have requested here. However, if we wanted to use a different estimator, we would simply replace ml in the estimator= command for our method of choice. In terms of specifying the factor structure, the major difference between this multilevel model and the single-level CFA models that we fit in Chapter 9 is that in this case, we must define the latent structure at both levels of the data. Note that the structure is hypothesized to be the same but the factors are differentiated as to whether they appear at level 1 (f1w, f2w, and f3w) or level 2 (f1b, f2b, and f3b).

The output containing the fit statistics produced by the resulting model appears below.

```
MODEL FIT INFORMATION

Number of Free Parameters                        66

Loglikelihood

        H0 Value                          -37566.922
        H1 Value                          -37527.170

Information Criteria

        Akaike (AIC)                       75265.845
        Bayesian (BIC)                     75635.504
        Sample-Size Adjusted BIC           75425.819
          (n* = (n + 2) / 24)

Chi-Square Test of Model Fit

        Value                                 79.504
        Degrees of Freedom                       102
        P-Value                               0.9516
```

```
RMSEA (Root Mean Square Error Of Approximation)

          Estimate                      0.000

CFI/TLI

          CFI                           1.000
          TLI                           1.005

Chi-Square Test of Model Fit for the Baseline Model

          Value                      5654.697
          Degrees of Freedom              132
          P-Value                      0.0000

SRMR (Standardized Root Mean Square Residual)

          Value for Within             0.014
          Value for Between            0.078
```

For the current example, overall fit looks very good using common guidelines (Kline, 2016), given that the Chi-square goodness of fit test is not statistically significant, the Root Mean Square Error Of Approximation (RMSEA) is 0, and the CFI and Tucker-Lewis Index (TLI) exceed 0.95. We have already seen from our prior analyses that the CFI values at levels 1 and 2 were both quite high, indicating good fit in each case. As noted, the SRMR provides information about fit of the model at both levels, and from these results, we can conclude that fit at level 1 is quite good (SRMR = 0.014) and that it is acceptable at level 2 (SRMR = 0.078). The fact that fit for the level 2 portion of the model is not as good as that for level 1 is not completely surprising, given that the number of level 2 units (60) is much smaller than the number of level 1 units (2000). Indeed, researchers working with multilevel CFA models must be cognizant of the level 2 sample size in particular, because, in many applications, it may be too small for the accurate estimation of model parameters.

Given that the model provides good fit to the data, we can next proceed to interpretation of the model parameters, in particular the factor loadings. The standardized estimates appear below for both levels of the data.

```
STANDARDIZED MODEL RESULTS


STDYX Standardization
```

		Estimate	S.E.	Est./S.E.	Two-Tailed P-Value

Within Level

F1W	BY				
Y1		0.727	0.014	51.088	0.000
Y2		0.707	0.015	48.117	0.000
Y3		0.696	0.015	46.367	0.000
Y4		0.718	0.014	49.868	0.000

F2W	BY				
Y5		0.689	0.018	38.429	0.000
Y6		0.563	0.020	28.051	0.000
Y7		0.531	0.021	25.613	0.000
Y8		0.547	0.020	26.903	0.000

F3W	BY				
Y9		0.695	0.029	24.238	0.000
Y10		0.347	0.027	12.907	0.000
Y11		0.290	0.028	10.537	0.000
Y12		0.458	0.027	17.261	0.000

F2W	WITH				
F1W		0.618	0.023	27.251	0.000

F3W	WITH				
F1W		0.232	0.032	7.221	0.000
F2W		0.535	0.032	16.747	0.000

Variances					
F1W		1.000	0.000	999.000	999.000
F2W		1.000	0.000	999.000	999.000
F3W		1.000	0.000	999.000	999.000

Residual Variances					
Y1		0.472	0.021	22.839	0.000
Y2		0.500	0.021	24.011	0.000
Y3		0.516	0.021	24.679	0.000
Y4		0.484	0.021	23.398	0.000
Y5		0.526	0.025	21.282	0.000
Y6		0.684	0.023	30.293	0.000
Y7		0.718	0.022	32.606	0.000
Y8		0.700	0.022	31.446	0.000
Y9		0.517	0.040	12.968	0.000
Y10		0.879	0.019	47.058	0.000
Y11		0.916	0.016	57.432	0.000
Y12		0.790	0.024	32.563	0.000

Between Level

F1B	BY				
Y1		1.000	0.013	74.618	0.000
Y2		0.999	0.011	88.176	0.000
Y3		0.998	0.011	92.809	0.000
Y4		1.000	0.014	73.635	0.000
F2B	BY				
Y5		0.999	0.011	90.751	0.000
Y6		0.999	0.017	59.938	0.000
Y7		0.971	0.021	46.393	0.000
Y8		1.000	0.023	44.145	0.000
F3B	BY				
Y9		1.007	0.017	59.996	0.000
Y10		0.980	0.038	25.941	0.000
Y11		0.909	0.099	9.211	0.000
Y12		0.982	0.023	41.867	0.000
F2B	WITH				
F1B		0.454	0.112	4.039	0.000
F3B	WITH				
F1B		0.782	0.063	12.513	0.000
F2B		0.650	0.083	7.830	0.000
Intercepts					
Y1		0.199	0.142	1.408	0.159
Y2		0.180	0.140	1.287	0.198
Y3		0.253	0.143	1.771	0.077
Y4		0.185	0.141	1.318	0.188
Y5		-0.040	0.136	-0.294	0.769
Y6		-0.029	0.141	-0.206	0.837
Y7		-0.050	0.140	-0.357	0.721
Y8		-0.004	0.140	-0.029	0.977
Y9		0.033	0.137	0.239	0.811
Y10		0.087	0.152	0.574	0.566
Y11		-0.018	0.182	-0.096	0.923
Y12		0.126	0.140	0.899	0.369
Variances					
F1B		1.000	0.000	999.000	999.000
F2B		1.000	0.000	999.000	999.000
F3B		1.000	0.000	999.000	999.000
Residual Variances					
Y1		0.001	0.027	0.027	0.979
Y2		0.002	0.023	0.081	0.935

Y3	0.003	0.021	0.141	0.888
Y4	0.001	0.027	0.024	0.981
Y5	0.002	0.022	0.085	0.933
Y6	0.002	0.033	0.059	0.953
Y7	0.057	0.041	1.406	0.160
Y8	0.001	0.045	0.012	0.991
Y9	-0.014	999.000	999.000	999.000
Y10	0.039	0.074	0.522	0.602
Y11	0.174	0.179	0.971	0.331
Y12	0.036	0.046	0.783	0.434

First, we can focus on the factor loadings at each level. An assumption is made that the covariance structure is equivalent across clusters at level 2. In this example, this would mean that we assume that the covariance matrix for the quality of life subscales is the same from one treatment center to the next. Thus, there exists only one set of factor loadings at level 1. At this level, the factor loadings are all statistically significant, meaning that they are different from 0; that is, each of the indicators is related to the latent variable with which it was hypothesized to measure. In addition, it appears that, descriptively, the loadings for Y10, Y11, and Y12 were somewhat smaller than the loadings for the other indicators in the model. At level 1, all of the factors were significantly correlated with one another, with the smallest correlation being between f1w and f3w (0.232) and the largest one being between f1w and f2w (0.618).

The between-level results reflect the latent structure among center-level means of the subscales. In other words, does the hypothesized three-factor structure hold when the indicators are the means of the quality of life sub-scales for each of the centers? Based on the results above, the answer would appear to be yes. All of the standardized loadings are near 1 and all are sta-tistically significant. It should be noted that the level 2 residual variances for these indicators were very small, and in one case, these were even negative. These low values are a consequence of trying to estimate the factor structure by using a relatively small level 2 sample ($N = 60$), combined with the strong underlying factor structure, which results in very large factor loadings, and resultant small error variances at level 2.

Estimating the Proportion of Variance Associated with Each Level of the Data

In order to estimate the amount of variance in the observed indicators that is due to each level of the data, we must specify the same latent structure at each level of the data and must constrain the factor loadings to be equivalent at both levels (Mehta and Neale, 2005). In other words, the loading for subscale 2 at level 1 must be equal to the loading for subscale 2 at level 2. This can be done easily in Mplus, as seen in the following program: `model_10.2_constrained.inp`.

```
TITLE: Model 10.2  two level CFA with
   loadings constrained to be equal
   at levels 1 and 2

DATA: FILE IS mlm_sem.dat;

VARIABLE: NAMES ARE y1-y12 center;
USEVARIABLES = y1-y12;
CLUSTER = center;

ANALYSIS: TYPE = TWOLEVEL;
estimator=ml;

MODEL:

   %WITHIN%
   f1w BY y1;
   f1w by y2(1);
   f1w by y3(2);
   f1w by y4(3);
   f2w by y5;
   f2w by y6(4);
   f2w by y7(5);
   f2w by y8(6);
   f3w by y9;
   f3w by y10(7);
   f3w by y11(8);
   f3w by y12(9);


   %BETWEEN%
   f1b BY y1;
   f1b by y2(1);
   f1b by y3(2);
   f1b by y4(3);
   f2b by y5;
   f2b by y6(4);
   f2b by y7(5);
   f2b by y8(6);
   f3b by y9;
   f3b by y10(7);
   f3b by y11(8);
   f3b by y12(9);

output: standardized;
```

The only difference between this model and Model 10.2 is that here, we have constrained the factor loadings to be equal to one across levels. We do this by using a number within (). Thus, the loading for y2 with f1w is constrained to

be equal to the loading for y2 with f1b because of the (2) on each. We can
see this equality of the loadings in the resulting output.

MODEL RESULTS

		Estimate	S.E.	Est./S.E.	Two-Tailed P-Value
Within Level					
F1W	BY				
Y1		1.000	0.000	999.000	999.000
Y2		0.998	0.030	33.038	0.000
Y3		0.948	0.029	32.296	0.000
Y4		1.002	0.030	33.337	0.000
F2W	BY				
Y5		1.000	0.000	999.000	999.000
Y6		0.702	0.027	26.200	0.000
Y7		0.650	0.027	23.855	0.000
Y8		0.679	0.026	25.847	0.000
F3W	BY				
Y9		1.000	0.000	999.000	999.000
Y10		0.407	0.026	15.360	0.000
Y11		0.291	0.026	11.394	0.000
Y12		0.603	0.034	17.526	0.000
Between Level					
F1B	BY				
Y1		1.000	0.000	999.000	999.000
Y2		0.998	0.030	33.038	0.000
Y3		0.948	0.029	32.296	0.000
Y4		1.002	0.030	33.337	0.000
F2B	BY				
Y5		1.000	0.000	999.000	999.000
Y6		0.702	0.027	26.200	0.000
Y7		0.650	0.027	23.855	0.000
Y8		0.679	0.026	25.847	0.000
F3B	BY				
Y9		1.000	0.000	999.000	999.000
Y10		0.407	0.026	15.360	0.000
Y11		0.291	0.026	11.394	0.000
Y12		0.603	0.034	17.526	0.000

In order to obtain the ICC for each factor, we would employ Equation 10.7
above, using the factor variances resulting from this constrained model.
Those values appear below.

```
Within Level
Variances
    F1W                 1.043        0.057        18.361        0.000
    F2W                 0.943        0.056        16.729        0.000
    F3W                 0.846        0.067        12.561        0.000
Between Level
Variances
    F1B                 0.378        0.078         4.864        0.000
    F2B                 0.542        0.109         4.994        0.000
    F3B                 0.492        0.098         5.009        0.000
```

The ICC (proportion of the latent variance accounted for at level 2) for factor 1 is then calculated as

$$\text{ICC}_{F1} = \frac{0.378}{0.378 + 1.043} = \frac{0.378}{1.421} = 0.266$$

Therefore, we can conclude that approximately 27% of the variation in the first latent variable is due to between-treatment center variance. Similar calculations for the other two factors yielded ICCs of 0.365 and 0.368, respectively.

Multilevel Structural Equation Modeling

In Chapter 9, we discussed structural equation models (SEMs), which allow us to relate latent variables to one another in much the same way that we do with observed variables using regression models. Moreover, just as we can extend single-level regression models to the multilevel context, so can we employ SEM in the multilevel context. Indeed, many of the key ideas that we described in Chapters 3 and 4 for multilevel regression models can be applied in the latent variable context with multilevel SEM. In this section, we will describe the basics of fitting multilevel SEMs by using Mplus.

The single-level SEM was written as

$$\eta = B\gamma + \zeta \tag{10.10}$$

where:

η is endogenous latent variable(s)

γ is exogenous latent variable(s)

B is coefficient linking the endogenous and exogenous variables with one another

ζ is random error with a mean of 0 and variance of ϕ

This model can easily be extended to the multilevel framework through specifications of the model at levels 1 and 2. The level 1 model appears in Equation 10.11.

$$\eta_{\text{wic}} = B_W \gamma_{\text{wic}} + \zeta_{\text{wic}} \tag{10.11}$$

where:

η_{wic} is endogenous latent variable(s) for subject i in cluster c

γ_{wic} is exogenous latent variable(s) for subject i in cluster c

B_W is within-cluster coefficient linking the two factors with one another

ζ_{wic} is random error for subject i in cluster c

Likewise, the level 2 portion of the model is expressed as

$$\eta_{Bc} = \alpha_c + B_B \gamma_{Bc} + \zeta_{Bc} \tag{10.12}$$

where:

α_c is vector of latent variable means for cluster c

η_{Bc} is endogenous latent variable(s) for cluster c

γ_{Bc} is esxogenous latent variable(s) for cluster c

B_B is between-cluster coefficient linking the two factors with one another

ζ_{Bc} is random error for cluster c

Fitting Multilevel SEM by Using Mplus

Fitting the multilevel SEM by using Mplus is very similar to fitting the multilevel factor models that we examined in the prior section of this chapter. Indeed, a preliminary step to estimating an SEM is that we assess the structure of the latent variables first to ensure that the individual measurement models fit the data, much as we did in the single-level case (see Mplus input files for Models 10.2a–10.2c). Once the suitability of the measurement models has been established, we can then turn our attention to the structural model itself. We will continue with the quality of life instrument that was introduced in the multilevel factor analysis section. In this case, we will treat factor 1 as the dependent (endogenous) variable and factors 2 and 3 as the exogenous variables. The program for fitting this model is in the file model_10.3.inp.

```
TITLE: Model 10.3  two level SEM

DATA: FILE IS mlm_sem.dat;

VARIABLE: NAMES ARE y1-y12 center;
USEVARIABLES = y1-y12;
CLUSTER = center;

ANALYSIS: TYPE = TWOLEVEL;
estimator=ml;
```

```
MODEL:

    %WITHIN%
    f1w BY y1-y4;
    f2w by y5-y8;
    f3w by y9-y12;
    f1w on f2w f3w;

    %BETWEEN%
    f1b BY y1-y4;
    f2b by y5-y8;
    f3b by y9-y12;
    y1@0;
    y5@0;
    y9@0;

    f1b on f2b f3b;
output: standardized;
```

This program resembles those used in the CFA examples above, with the primary addition being the inclusion of the structural portion of the model at each level, that is, f1w on f2w f3w; and f1b on f2b f3b;.

In examining the output, we will focus first on the ICCs for each of the observed variables.

```
Estimated Intraclass Correlations for the Y Variables
```

Variable	Intraclass Correlation	Variable	Intraclass Correlation	Variable	Intraclass Correlation
Y1	0.150	Y2	0.165	Y3	0.151
Y4	0.162	Y5	0.223	Y6	0.142
Y7	0.147	Y8	0.151	Y9	0.197
Y10	0.079	Y11	0.034	Y12	0.158

We can interpret from these results that each of the subscales, with the exceptions of Y10 and Y11, exhibited a moderate amount of ICC or a proportion of variation associated with level 2. The model fit information appears next.

```
MODEL FIT INFORMATION

Number of Free Parameters                       63

Loglikelihood

        H0 Value                          -37567.226
        H1 Value                          -37527.170

Information Criteria
        Akaike (AIC)                       75260.452
        Bayesian (BIC)                     75613.309
```

```
        Sample-Size Adjusted BIC              75413.154
           (n* = (n + 2) / 24)
```

Chi-Square Test of Model Fit

```
        Value                                    80.111
        Degrees of Freedom                          105
        P-Value                                  0.9663
```

RMSEA (Root Mean Square Error Of Approximation)

```
        Estimate                                  0.000
```

CFI/TLI

```
        CFI                                       1.000
        TLI                                       1.006
```

Chi-Square Test of Model Fit for the Baseline Model

```
        Value                                  5654.697
        Degrees of Freedom                          132
        P-Value                                  0.0000
```

SRMR (Standardized Root Mean Square Residual)

```
        Value for Within                          0.014
        Value for Between                         0.078
```

These results are very similar to those of the full CFA model (Model 10.2) and indicate very good fit. Again, it is important to keep in mind that except for the SRMR, these indices are dominated by the level 1 portion of the model, because its sample size is so much larger (2000 vs. 60).

Finally, we will consider the model parameter estimates. In particular, we are interested in the structural coefficients, as opposed to the factor loadings, which were our primary concern in the context of CFA. Indeed, once we have established that the individual measurement models fit the data, we are typically not very interested in the factor loading estimates. For this reason, we include only the standardized structural values for the model below.

STANDARDIZED MODEL RESULTS

STDYX Standardization

	Estimate	S.E.	Est./S.E.	Two-Tailed P-Value
Within Level				
F1W ON				
F2W	0.692	0.036	19.435	0.000
F3W	-0.138	0.042	-3.246	0.001
F3W WITH				
F2W	0.534	0.032	16.743	0.000
Between Level				
F1B ON				
F2B	-0.096	0.133	-0.723	0.470
F3B	0.845	0.109	7.761	0.000
F3B WITH				
F2B	0.650	0.083	7.800	0.000

Based on these results, we can conclude that at level 1, latent variable 2 has a statistically significant positive relationship with latent variable 1, whereas latent variable 3 has a significantly negative relationship with latent variable 1. In other words, individuals with higher values on the second factor also have higher values on the first factor, whereas those with higher values on the third factor have lower values on the second factor. At level 2, factor 2 is not related to factor 1 ($p = 0.470$), but factor 3 has a statistically significant positive association with factor 1. This means that when the average factor 3 score at a center was higher, the average factor 1 score was also higher.

Random Coefficient SEM

We saw in Chapter 3 that it is possible to fit a multilevel model with a random coefficient, relating the independent and dependent variables with one another. This random coefficient reflects the extent to which the relationships between the independent and dependent variables vary across the level 2 units. Similarly, with SEM, we can model random coefficients for one or more of the coefficients relating the endogenous and exogenous factors in Equations 10.11 and 10.12. The Mplus syntax necessary to do this is very similar in structure to what we saw with the random coefficient models in Chapter 3, involving the creation of a specific term for the coefficient itself at level 1 (within). In order to assess whether this coefficient varies across the level 2 clusters, we simply estimate the variance of this coefficient term and test it for statistical significance. The null hypothesis for this test is that there is no variation in the coefficients across the level 2 units. Continuing with the example involving the quality of life measures for patients at various cancer treatment centers, we are interested in determining whether the relationship between factor 2 and factor 1 varies across level 2 units. Because of the computation burden placed by this model, we are fitting the relationship only between these two factors and ignoring factor 3 altogether. The Mplus syntax for this analysis is contained in `model_10.3_random_slope.inp`.

```
TITLE: Model 10.3  two level SEM with random slope

DATA: FILE IS mlm_sem.dat;

VARIABLE: NAMES ARE y1-y12 center;
USEVARIABLES = y1-y8;
CLUSTER = center;

ANALYSIS: TYPE = TWOLEVEL random;
algorithm=integration;
integration=10;
estimator=ml;

MODEL:

    %WITHIN%
    f1w BY y1-y4;
    f2w by y5-y8;

    f1w on f2w;
    s | f1w on f2w;

    %BETWEEN%
    f1b BY y1-y4;
    f2b by y5-y8;
    y1@0;
    y5@0;

    f1b on f2b;

    s;

output: standardized;
```

There are several new elements to this program that we need to examine. First, when fitting the random coefficient SEM, we must use numerical integration, which we specify with the `algorithm=integration` subcommand under `ANALYSIS`. The `integration=10` subcommand specifies the number of integration points per dimension that should be used. Increasing this number may yield more precise estimates but will also increase the computational burden of doing the analysis. Next, we create the random coefficient term by using s | f1w on f2w; in the within part of the model, much as we did for observed variable models in earlier chapters of the book. Finally, in order to obtain the variance of this term, we specify it in the between portion of the model code as s;. The relevant output appears below.

Between Level

	Estimate	S.E.	Est./S.E.	Two-Tailed P-Value
Means				
S	0.658	0.042	15.648	0.000
Variances				
F2B	0.553	0.111	4.960	0.000
S	0.015	0.011	1.317	0.188

The sample estimate for the mean of the coefficient linking f1w and f2w was 0.658 and was statistically significant. The variance of the coefficient was not statistically significant ($p = 0.188$). In other words, there was no variation across centers in the nature of the relationship between factor 1 and factor 2, and overall, this relationship was statistically significant, with a sample estimate of 0.658.

Multilevel Growth Curve Models

One of the paradigms that we examined in Chapter 9 involved the use of modeling longitudinal data using growth curve models (GCMs). With such models, we were interested in characterizing change in a measurement that was made at multiple points in time. As we saw in Chapter 6, this type of problem can also be addressed by using multilevel models, whereby the individual measurement occasions are considered to be at level 1 and the subjects on whom the measurements are made are considered to be at level 2. There are advantages to both approaches, with greater flexibility in terms of the form of the model that can be fit being the primary advantage of the GCM approach. In this chapter, we extend the standard GCM to include a second (or third, depending on your viewpoint) level of data. As an example, consider the case where 700 students from 60 different elementary schools were followed for 6 years. At the end of each year, they were given a mathematics achievement test. The researchers are interested in determining whether achievement changed over time, and if so, how it changed. This problem is very much akin to the GCM described in Chapter 9, except that an additional layer of data has been added in the form of school. Therefore, we now consider change in scores over time both for students (level 1) and for schools (level 2). The Mplus program for fitting this model is presented below and appears in the file model_10.4.inp.

```
TITLE: Model 10.4 Two level GCM
DATA:  FILE IS mlm_gcm.dat;
VARIABLE:      NAMES ARE y1-y6 school;
       CLUSTER = school;

ANALYSIS:      TYPE = TWOLEVEL;

MODEL:
       %WITHIN%
       iw sw | y1@0 y2@1 y3@2 y4@3 y5@4 y6@5;
       y1-y6 (1);

       %BETWEEN%
       ib sb | y1@0 y2@1 y3@2 y4@3 y5@4 y6@5;
       y1-y6@0;

output: Standardized;
```

In many respects, the structure of the program is similar to that of the single-level GCM, with the exception that we now have two levels for the model. For model identification purposes, we must constrain the variances of the measurements to be equal at level 1 (y1-y6 (1) ;) and to be 0 at level 2 (y1-y6@0 ;). The resulting output appears below.

MODEL RESULTS

	Estimate	S.E.	Est./S.E.	Two-Tailed P-Value
Within Level				
IW				
Y1	1.000	0.000	999.000	999.000
Y2	1.000	0.000	999.000	999.000
Y3	1.000	0.000	999.000	999.000
Y4	1.000	0.000	999.000	999.000
Y5	1.000	0.000	999.000	999.000
Y6	1.000	0.000	999.000	999.000
SW				
Y1	0.000	0.000	999.000	999.000
Y2	1.000	0.000	999.000	999.000
Y3	2.000	0.000	999.000	999.000
Y4	3.000	0.000	999.000	999.000
Y5	4.000	0.000	999.000	999.000
Y6	5.000	0.000	999.000	999.000
SW WITH				
IW	-0.014	0.038	-0.360	0.719

Variances
IW	1.106	0.063	17.438	0.000
SW	0.730	0.039	18.507	0.000

Residual Variances
Y1	0.498	0.014	35.637	0.000
Y2	0.498	0.014	35.637	0.000
Y3	0.498	0.014	35.637	0.000
Y4	0.498	0.014	35.637	0.000
Y5	0.498	0.014	35.637	0.000
Y6	0.498	0.014	35.637	0.000

Between Level

IB |
Y1	1.000	0.000	999.000	999.000
Y2	1.000	0.000	999.000	999.000
Y3	1.000	0.000	999.000	999.000
Y4	1.000	0.000	999.000	999.000
Y5	1.000	0.000	999.000	999.000
Y6	1.000	0.000	999.000	999.000

SB |
Y1	0.000	0.000	999.000	999.000
Y2	1.000	0.000	999.000	999.000
Y3	2.000	0.000	999.000	999.000
Y4	3.000	0.000	999.000	999.000
Y5	4.000	0.000	999.000	999.000
Y6	5.000	0.000	999.000	999.000

SB   WITH
IB	0.046	0.055	0.841	0.400

Means
IB	1.582	0.078	20.244	0.000
SB	2.123	0.092	23.172	0.000

Intercepts
Y1	0.000	0.000	999.000	999.000
Y2	0.000	0.000	999.000	999.000
Y3	0.000	0.000	999.000	999.000
Y4	0.000	0.000	999.000	999.000
Y5	0.000	0.000	999.000	999.000
Y6	0.000	0.000	999.000	999.000

Variances
IB	0.228	0.060	3.815	0.000
SB	0.419	0.088	4.773	0.000

```
Residual Variances
    Y1                    0.000         0.000      999.000     999.000
    Y2                    0.000         0.000      999.000     999.000
    Y3                    0.000         0.000      999.000     999.000
    Y4                    0.000         0.000      999.000     999.000
    Y5                    0.000         0.000      999.000     999.000
    Y6                    0.000         0.000      999.000     999.000
```

These results indicate that the average growth across schools was 2.123 per year and the mean starting score was 1.582, both of which were statistically significantly different from 0. In addition, there was not a significant relationship between the starting score and growth over time, with the sample covariance between these two parameters estimated to be 0.046. Finally, there was statistically significant variation in the degree of growth among individuals within schools ($\hat{\sigma}_{SW}^2 = 0.730$), as well as there was significant variance in the starting values within schools ($\hat{\sigma}_{IW}^2 = 0.730$). In other words, not all of the students started at the same level and not all of them had scores change at the same rate over time.

As with single-level GCMs, researchers can include variables to explain the growth parameter at levels 1 and 2. In addition, random coefficient models can be fit, allowing for the relationship of these covariates to the change parameter to vary across the level 2 units. Finally, we need to say a word about the terminology that we have used in this section of the chapter. Standard GCMs can be thought of as two-level models in the sense that the measurements taken at each point in time are at level 1 and the individuals on whom the measurements are made are at level 2. Thus, when we add in a third level of data, such as school in the prior example, we are fitting what amounts to a three-level model. In some texts and manuscripts, authors use this terminology to describe such models. However, we have elected to refer to the subjects on whom measurements are made as level 1 units and the clustering variable (e.g., school) as level 2 units. This decision was made in order to keep the language consistent across examples (i.e., level 1 is subject and level 2 is clustering unit). However, we recognize that strictly speaking, the GCM paradigm corresponds to a three-level model.

Multilevel Item Response Theory (IRT) Models

Multilevel modeling in the context of item response theory (IRT) is very similar in spirit to multilevel modeling for FA models, which may well be due to the close association between these two modeling paradigms (McDonald, 1999). Therefore, many of the concepts that we have described earlier for FA models, particularly with regard to the presence of indicator-specific parameter estimates at levels 1 and 2, will also apply in the context of IRT. Recall from Chapter 9 that for dichotomous data, the 2-parameter logistic (2PL) model takes the form.

$$P\left(x_j = 1 \mid \theta_i, a_j, b_j\right) = \frac{e^{a_j\left(\theta_i - b_j\right)}}{1 + e^{a_j\left(\theta_i - b_j\right)}} \qquad (10.13)$$

where:
θ_i is latent trait being measured for subject i
b_j is difficulty (location) for item j
a_j is discrimination for item j

In the multilevel context, we need to include information about level 2 in the model. As described by Kamata and Vaughn (2011), the multilevel 2PL model can be expressed as

$$P\left(x_j = 1 \mid \theta_{ic}, a_j, b_j\right) = \frac{e^{a_j\left(\theta_{ic} - b_j\right)}}{1 + e^{a_j\left(\theta_{ic} - b_j\right)}} \qquad (10.14)$$

The terms in Equation 10.14 are identical to those in Equation 10.13, with the exception of the latent trait being measured. In the single-level analysis, θ_i is simply the latent trait for individual i. However, in the multilevel context, the latent trait becomes θ_{ic} or the latent trait for individual i in cluster c. As described by Kamata and Vaughn, θ_{ic} can be decomposed into two components, ξ_c, which is the mean level of the latent trait for cluster c, and ζ_{ic}, which is the deviation from this mean by individual i in cluster c. In addition, researchers can include covariates of the latent trait in this model, and the covariates can occur at level 1, level 2, or both.

Fitting a Multilevel IRT Model by Using Mplus

The example that we will use to demonstrate the fitting of multilevel IRT models by using Mplus involves 30 reading test items (scored correct/incorrect) that were given to 1600 fifth-grade children across 95 schools. The primary goal of the analysis is to estimate item difficulty and discrimination parameters for the test items. The following Mplus program, model_10.5.inp, was used to fit the 2PL IRT model.

```
TITLE:     Model 10.5  Two level 2PL model for
        dichotomous data
DATA:  FILE IS mlm_irt.dat;
VARIABLE:     NAMES ARE item1-item30 school;
        CATEGORICAL = item1-item30;
        CLUSTER = school;
```

```
ANALYSIS:     TYPE = TWOLEVEL;
              estimator=mlr;
MODEL:
       %WITHIN%
       thetaw BY item1* (0)
                 item2  (1)
                 item3  (2)
                 item4  (3)
                 item5  (4)
                 item6  (5)
                 item7  (6)
                 item8  (7)
                 item9  (8)
                 item10 (9)
                 item11 (10)
                 item12 (11)
                 item13 (12)
                 item14 (13)
                 item15 (14)
                 item16 (15)
                 item17 (16)
                 item18 (17)
                 item19 (18)
                 item20 (19)
                 item21 (20)
                 item22 (21)
                 item23 (22)
                 item24 (23)
                 item25 (24)
                 item26 (25)
                 item27 (26)
                 item28 (27)
                 item29 (28)
                 item30 (29);

                 thetaw@1;
                 [thetaw@0];


       %BETWEEN%
       thetab BY item1*  (0)
                 item2  (1)
                 item3  (2)
                 item4  (3)
                 item5  (4)
                 item6  (5)
                 item7  (6)
```

```
             item8  (7)
             item9  (8)
             item10 (9)
             item11 (10)
             item12 (11)
             item13 (12)
             item14 (13)
             item15 (14)
             item16 (15)
             item17 (16)
             item18 (17)
             item19 (18)
             item20 (19)
             item21 (20)
             item22 (21)
             item23 (22)
             item24 (23)
             item25 (24)
             item26 (25)
             item27 (26)
             item28 (27)
             item29 (28)
             item30 (29);

             [thetab@0];
OUTPUT:      TECH1 TECH8;
```

The basic framework of this program looks very much like what we used for the CFA modeling earlier in this chapter, with some adjustments. In particular, given Equation 10.14, we assume that there is one set of item parameters transcending levels 1 and 2. As we saw with the CFA example, it is possible to constrain model parameters to be equal by associating them with a number between (). Thus, in this case, we are constraining the discrimination parameter estimates for each item to be equal across levels 1 and 2; for example, the discrimination estimate for item 2 is equal at levels 1 and 2 because of the placement of (1) after each. In addition, we free the discrimination parameter value for item 1 by including * at both the within and between levels. Finally, we identify the model by setting the variance of the level 1 latent trait value to be 1 (thetaw@1;). The means at both levels are set equal to 0 (e.g., [thetab@0];). Selected portions of the output resulting from this analysis appear below.

First, we examine the model fit information. If we were interested in fitting alternative models to the data, such as 1-parameter logistic (1PL) or Rasch, we could compare the fit of the various models by using the information indices.

```
MODEL FIT INFORMATION

Number of Free Parameters                        61
Loglikelihood

        H0 Value                          -30654.097
        H0 Scaling Correction Factor        1.0032
          for MLR

Information Criteria

        Akaike (AIC)                       61430.193
        Bayesian (BIC)                     61758.236
        Sample-Size Adjusted BIC           61564.451
          (n* = (n + 2) / 24)
```

The model parameter estimates appear next.

```
MODEL RESULTS

                                                        Two-Tailed
                       Estimate     S.E.    Est./S.E.    P-Value


Within Level

THETAW    BY
    ITEM1               0.957      0.076     12.578       0.000
    ITEM2               0.736      0.065     11.242       0.000
    ITEM3               0.850      0.068     12.475       0.000
    ITEM4               0.743      0.062     12.036       0.000
    ITEM5               0.882      0.078     11.355       0.000
    ITEM6               0.855      0.068     12.606       0.000
    ITEM7               0.918      0.067     13.713       0.000
    ITEM8               0.856      0.057     15.140       0.000
    ITEM9               0.802      0.066     12.160       0.000
    ITEM10              0.759      0.071     10.634       0.000
    ITEM11              0.720      0.058     12.334       0.000
    ITEM12              0.801      0.062     12.928       0.000
    ITEM13              0.867      0.067     12.844       0.000
    ITEM14              0.844      0.066     12.867       0.000
    ITEM15              0.754      0.058     13.072       0.000
    ITEM16              0.684      0.060     11.363       0.000
    ITEM17              0.820      0.067     12.268       0.000
    ITEM18              0.758      0.067     11.383       0.000
    ITEM19              0.840      0.073     11.514       0.000
    ITEM20              0.936      0.067     14.023       0.000
    ITEM21              0.799      0.060     13.371       0.000
    ITEM22              0.737      0.058     12.750       0.000
```

ITEM23	0.837	0.066	12.623	0.000
ITEM24	0.873	0.062	14.156	0.000
ITEM25	0.795	0.071	11.193	0.000
ITEM26	0.783	0.074	10.631	0.000
ITEM27	0.795	0.066	12.035	0.000
ITEM28	0.818	0.068	12.077	0.000
ITEM29	0.838	0.065	12.831	0.000
ITEM30	0.818	0.061	13.514	0.000

Means
THETAW	0.000	0.000	999.000	999.000

Variances
THETAW	1.000	0.000	999.000	999.000

Between Level

THETAB   BY
ITEM1	0.957	0.076	12.578	0.000
ITEM2	0.736	0.065	11.242	0.000
ITEM3	0.850	0.068	12.475	0.000
ITEM4	0.743	0.062	12.036	0.000
ITEM5	0.882	0.078	11.355	0.000
ITEM6	0.855	0.068	12.606	0.000
ITEM7	0.918	0.067	13.713	0.000
ITEM8	0.856	0.057	15.140	0.000
ITEM9	0.802	0.066	12.160	0.000
ITEM10	0.759	0.071	10.634	0.000
ITEM11	0.720	0.058	12.334	0.000
ITEM12	0.801	0.062	12.928	0.000
ITEM13	0.867	0.067	12.844	0.000
ITEM14	0.844	0.066	12.867	0.000
ITEM15	0.754	0.058	13.072	0.000
ITEM16	0.684	0.060	11.363	0.000
ITEM17	0.820	0.067	12.268	0.000
ITEM18	0.758	0.067	11.383	0.000
ITEM19	0.840	0.073	11.514	0.000
ITEM20	0.936	0.067	14.023	0.000
ITEM21	0.799	0.060	13.371	0.000
ITEM22	0.737	0.058	12.750	0.000
ITEM23	0.837	0.066	12.623	0.000
ITEM24	0.873	0.062	14.156	0.000
ITEM25	0.795	0.071	11.193	0.000
ITEM26	0.783	0.074	10.631	0.000
ITEM27	0.795	0.066	12.035	0.000
ITEM28	0.818	0.068	12.077	0.000
ITEM29	0.838	0.065	12.831	0.000
ITEM30	0.818	0.061	13.514	0.000

Means
    THETAB   0.000   0.000   999.000   999.000

Thresholds
ITEM1$1	-0.004	0.076	-0.055	0.956
ITEM2$1	0.013	0.062	0.216	0.829
ITEM3$1	0.087	0.076	1.149	0.250
ITEM4$1	0.097	0.070	1.397	0.162
ITEM5$1	0.069	0.067	1.043	0.297
ITEM6$1	0.150	0.073	2.060	0.039
ITEM7$1	0.033	0.079	0.420	0.675
ITEM8$1	0.063	0.067	0.939	0.348
ITEM9$1	0.116	0.077	1.510	0.131
ITEM10$1	0.025	0.065	0.381	0.703
ITEM11$1	0.094	0.074	1.264	0.206
ITEM12$1	0.151	0.074	2.038	0.042
ITEM13$1	0.145	0.075	1.927	0.054
ITEM14$1	0.006	0.081	0.079	0.937
ITEM15$1	0.132	0.074	1.775	0.076
ITEM16$1	0.142	0.074	1.912	0.056
ITEM17$1	0.033	0.076	0.437	0.662
ITEM18$1	0.118	0.066	1.792	0.073
ITEM19$1	0.117	0.071	1.651	0.099
ITEM20$1	0.067	0.079	0.845	0.398
ITEM21$1	0.163	0.083	1.967	0.049
ITEM22$1	0.051	0.069	0.738	0.461
ITEM23$1	0.096	0.062	1.546	0.122
ITEM24$1	0.051	0.071	0.718	0.473
ITEM25$1	0.051	0.079	0.642	0.521
ITEM26$1	0.189	0.075	2.526	0.012
ITEM27$1	0.060	0.071	0.841	0.400
ITEM28$1	0.051	0.069	0.735	0.462
ITEM29$1	0.081	0.072	1.119	0.263
ITEM30$1	0.007	0.074	0.090	0.928

Variances
    THETAB   0.253   0.048   5.308   0.000

Unlike the case with single-level data, in the multilevel context, Mplus does not provide the item parameter estimates in the IRT scale. Thus, the results that we see here are based on the FA paradigm, with factor loadings corresponding to item discrimination parameters and thresholds corresponding to item difficulty values. However, given the equivalence of the 2PL and CFA for categorical indicators models, it is very easy to convert the parameters from one to those of the other (McDonald, 1999). In order to obtain IRT discrimination values from factor loadings, we would use Equation 10.15.

$$a_j = \frac{\lambda_j}{\sqrt{1-\lambda_j^2}} \qquad (10.15)$$

where λ_j is factor loading for item j.

The item difficulty estimates are calculated as

$$b_j = \frac{-\tau_j}{\sqrt{1-\lambda_j^2}} \qquad (10.16)$$

where τ_j is threshold for item j.

As an example, let us calculate the IRT parameter estimates for item 2, using the results in the Mplus output.

$$a_2 = \frac{0.736}{\sqrt{1-0.736^2}} = \frac{0.736}{0.678} = 1.09$$

$$b_2 = \frac{-(0.013)}{\sqrt{1-0.736^2}} = \frac{-0.013}{0.678} = 0.019$$

We can make similar calculations for the rest of the items in order to obtain the item parameter estimates in the IRT scale.

Multilevel Latent Class Models

In Chapter 9, we described mixture modeling, which involves investigation of the population for the presence of subgroups, or classes, associated with differential patterns on observed variables. LCA models are characterized by two parameters: (1) the probability of a randomly selected individual in the population being in a particular latent class, and (2) the probability of a member of a particular latent class yielding a particular set of response to the observed variables. We described this model in Equation 9.19 and present it again here in Equation 10.17.

$$\pi_{1,2,3\dots,jt}^{X_1 X_2 X_3 \dots X_j, Y} = \sum_t^T \pi_t^Y \Pi \pi_{jt}^{X_j|Y} \qquad (10.17)$$

where:

π_t^Y is probability that a randomly selected individual will be in class t of latent variable Y

$\pi_{jt}^{X_j|Y}$ is probability that a member of latent class t will provide a particular response to observed indicator j

An assumption underlying mixture models is that once we condition on latent class membership, responses to the observed variables are independent of one another (local independence). However, when data are collected in a multilevel framework, this assumption is not valid, leading to estimation problems (Vermunt, 2003). When we have multilevel data, researchers interested in conducting an LCA need to accommodate this additional structure. Multilevel LCA (MLCA) offers just such an approach for dealing with these nested data. There exist two broad paradigms for fitting MLCA models, one parametric and the other nonparametric. The parametric model expresses the probability of being in latent class t as

$$P\left(C_{ij} = t\right) = \frac{\exp\left(\gamma_0 + \beta_1 x_{ij} + \gamma_1 w_j + U_{0j}\right)}{1 + \exp\left(\gamma_0 + \beta_1 x_{ij} + \gamma_1 w_j + U_{0j}\right)} \tag{10.18}$$

where:

C_{ij} is latent class for individual i in cluster j

γ_0 is intercept at level 2

x_{ij} is indicator variables at level 1 for individual i in cluster j

β_1 is coefficients for level 1 indicators

w_j is indicator variables for cluster j

γ_1 is coefficients for level 2 indicators

U_{0j} is random deviation from overall population average for cluster j

This model accounts for cluster membership by allowing latent class intercepts to vary at level 2, meaning that the probability of membership in a level 1 latent class can vary at level 2 (Asparouhov and Muthén, 2008; Vermunt). The random intercept is assumed to be a normally distributed latent variable in the MLCA model and is allowed to vary across clusters. There are T-1 such random intercepts, where T is the number of level 1 latent classes. The parametric model can be computationally intense, because a threshold parameter must be estimated for each indicator variable. This model complexity can lead to convergence problems and/or improper solutions in some instances. Thus, an alternative formulation of this model has been suggested, involving the inclusion of a single common factor reflecting the indicator-specific cluster influence (Asparouhov and Muthén, 2008; Vermunt). Each indicator variable has a unique factor loading, thus reducing the dimensionality of the parametric approach from T-1 random intercepts to a single factor. Thus, fewer parameters need to be estimated.

An alternative to the parametric model is a nonparametric approach, in which separate latent class models are specified for levels 1 and 2, with the T-1 random means from the level 1 latent class solution serving as indicators of the latent classes at level 2. In addition, rather than assuming normality of the random intercepts at level 2, as in the parametric models, the non-parametric approach assumes a multinomial distribution of the level 2 latent classes, resulting in a computationally simpler model that is not tied to the normality assumption (Asparouhov and Muthén, 2008). These level 2 latent classes reflect differences in the probability of belonging to a specific level 1 latent class, so that clusters (e.g., schools) that contain individuals with similar probabilities for the level 1 latent classes will be grouped together. The nonparametric MLCA model takes the form:

$$P\left(C_{ij} = t \mid CB_j = m\right) = \frac{\exp\left(\gamma_{tm}\right)}{\sum_{r=1}^{T} \exp\left(\gamma_{rm}\right)} \tag{10.19}$$

where:
  CB_j is level 2 latent class membership for cluster j
  γ_{tm} is level 1 and level 2 indicators

A variation of this nonparametric model allows a level 2 factor on the latent class indicators, in much the same fashion as for the parametric model. Each of these models allows for inclusion of covariates at both levels, where level 1 covariates predict class membership for individuals and level 2 covariates predict the probability that a level 1 latent class will be in a specific level 2 class. These models are described graphically in Figure 10.1, below.

NOTE: There will be T-1 latent class means, where T = number of latent classes. Latent class 1 will always have a mean of 0.

Finch and French (2014) conducted a simulation study comparing the performance of the parametric and nonparametric models and found that the nonparametric approach generally resulted in more accurate recovery of the underlying latent structure of the data at both levels.

Estimating MLCA in Mplus

Given that the nonparametric model was shown to provide better latent class recovery in most situations, it will be the focus in terms of our application with Mplus. For this example, we will use data from the Programme for International Student Assessment (PISA). Specifically, we are interested in identifying latent classes of students based on their achievement and socioeconomic status (SES). For the current example, 167,767 students from 6143 schools and 20 nations were included in the analysis. In this case, we will treat school as the only clustering variable. While it is true that nation may be

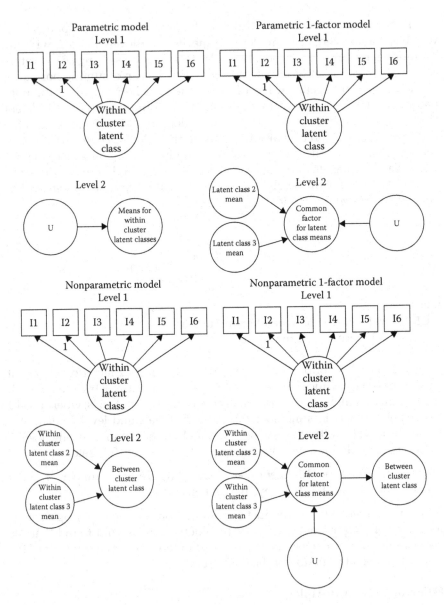

FIGURE 10.1
Multilevel latent class models.

an important source of within-group variation, with only 20 countries represented, parameter estimates at this level would likely not be very reliable. However, given the large number of schools, this would seem to be an excellent candidate to serve at level 2. The Mplus program used to fit the model for five level 1 and four level 2 latent classes appears below. This model was selected for demonstration here because, based on the information indices (i.e., AIC, BIC, and aBIC), it yielded the optimal fit from among a number of models that were fit to the data. It needs to be noted here that these models can take a nontrivial amount of time to run, which, in some cases, may exceed 2 hours. Therefore, the researcher will want to be judicious in her application of these models to large datasets and set aside sufficient time to fit all of the models that may be of interest.

```
TITLE:          Nonparametric multilevel LCA
DATA:           FILE IS mlm_lca.csv;
VARIABLE:       NAMES ARE COUNTRY StIDStd ESCS W_FSTUWT
   MATH READ SCIENCE school;
      USEVARIABLES = math read science escs;
      auxiliary=COUNTRY StIDStd W_FSTUWT;
      missing are all(9999);
      CLASSES = cb(4) cw(5);
   between=cb;
      WITHIN = math read science escs;
      CLUSTER = school;
ANALYSIS:       TYPE = TWOLEVEL MIXTURE;
MODEL:
      %WITHIN%
      %OVERALL%

      %BETWEEN%
      %OVERALL%
   cw on cb;

model cw:
      %WITHIN%
      %cw#1%
[math read science escs];
      %cw#2%
[math read science escs];
      %cw#3%
[math read science escs];
      %cw#4%
[math read science escs];
      %cw#5%
[math read science escs];
   savedata:
      file is resultscb4cw5.txt;
      save=cprob;
```

In this example, we are developing latent classes based on four variables, including scores on math, reading, and science achievement, as well as an index of SES (escs). We specify the number of between (4 for school) and within (5 for student) latent classes with cb(4) and cw(5), respectively. In addition, we must indicate to Mplus that cb is a between-subjects variable, whereas the four observed variables of interest are within subjects. We then indicate that cb is a predictor of cw (cw on cb), followed by the specific definition of how we would like for each of the level 1 (within) latent classes to be determined. If we were to reduce the number of level 1 classes to four, then the model cw: command would appear as:

```
     %cw#1%
[math read science escs];
     %cw#2%
[math read science escs];
     %cw#3%
[math read science escs];
     %cw#4%
[math read science escs];
```

Finally, in the event that we would like to save the predicted latent class membership and probabilities associated with each class for each individual, we can use the savedata command. The resulting class memberships, along with the variables used to create the classes and the variables listed in the auxiliary command that were not used in the LCA modeling, are all saved to a new file called resultscb4cw5.txt. We can then perform further follow-up analyses by using these results, such as comparing latent classes on other variables of interest and obtaining class-specific means for the variables used to create the latent classes, as well as others that might be of interest.

The (rather lengthy) relevant output produced by the program above appears below. First, we will examine the model fit information and class membership probabilities.

```
MODEL FIT INFORMATION

Number of Free Parameters                        43

Loglikelihood

          H0 Value                      -2915611.087
          H0 Scaling Correction Factor       5.1160
             for MLR

Information Criteria

          Akaike (AIC)                   5831308.174
          Bayesian (BIC)                 5831739.479
          Sample-Size Adjusted BIC       5831602.824
             (n* = (n + 2) / 24)
```

MODEL RESULTS USE THE LATENT CLASS VARIABLE ORDER

CB   CW

Latent Class Variable Patterns

CB Class	CW Class
1	1
1	2
1	3
1	4
1	5
2	1
2	2
2	3
2	4
2	5
3	1
3	2
3	3
3	4
3	5
4	1
4	2
4	3
4	4
4	5

FINAL CLASS COUNTS AND PROPORTIONS FOR THE LATENT CLASS PATTERNS
BASED ON ESTIMATED POSTERIOR PROBABILITIES

Latent Class
  Pattern

1	1	7362.70892	0.04389
1	2	71.45064	0.00043
1	3	3519.21466	0.02098
1	4	5713.37837	0.03405
1	5	729.57117	0.00435
2	1	236.14603	0.00141
2	2	10481.68073	0.06248
2	3	1776.76834	0.01059
2	4	22.03984	0.00013
2	5	8614.30120	0.05135
3	1	7394.47997	0.04407
3	2	9063.07760	0.05402
3	3	20771.95552	0.12381
3	4	926.02675	0.00552

```
3   5        25882.02449           0.15427
4   1        18909.69780           0.11271
4   2         2351.36078           0.01402
4   3        25385.74993           0.15131
4   4         4810.06696           0.02867
4   5        13750.30029           0.08196
```

FINAL CLASS COUNTS AND PROPORTIONS FOR EACH LATENT CLASS
VARIABLE
BASED ON ESTIMATED POSTERIOR PROBABILITIES

```
Latent Class
  Variable    Class

  CB            1      17396.32227         0.10369
                2      21130.93750         0.12595
                3      64037.56250         0.38169
                4      65207.17578         0.38867
  CW            1      33903.03125         0.20208
                2      21967.57031         0.13094
                3      51453.68750         0.30669
                4      11471.51270         0.06838
                5      48976.19922         0.29192
```

FINAL CLASS COUNTS AND PROPORTIONS FOR THE LATENT CLASS
PATTERNS
BASED ON THEIR MOST LIKELY LATENT CLASS PATTERN

Class Counts and Proportions

```
Latent Class
  Pattern

  1   1            7375            0.04396
  1   2              65            0.00039
  1   3            3434            0.02047
  1   4            5752            0.03428
  1   5             658            0.00392
  2   1             211            0.00126
  2   2           10524            0.06273
  2   3            1696            0.01011
  2   4              20            0.00012
  2   5            8657            0.05160
  3   1            7167            0.04272
  3   2            8907            0.05309
  3   3           20837            0.12420
  3   4             859            0.00512
  3   5           26402            0.15737
```

4	1	19079	0.11372
4	2	2190	0.01305
4	3	25755	0.15351
4	4	4736	0.02823
4	5	13448	0.08016

FINAL CLASS COUNTS AND PROPORTIONS FOR EACH LATENT CLASS
VARIABLE
BASED ON THEIR MOST LIKELY LATENT CLASS PATTERN

Latent Class Variable	Class		
CB	1	17284	0.10302
	2	21108	0.12581
	3	64172	0.38250
	4	65208	0.38867
CW	1	33832	0.20165
	2	21686	0.12926
	3	51722	0.30829
	4	11367	0.06775
	5	49165	0.29305

CLASSIFICATION QUALITY

Entropy   0.877

The largest latent class combination was level 2 class 3 with level 1 class 5, which included 26,402 individuals. The smallest class combination was level 2 class 2 with level 1 class 4, with only 20 people. When collapsing across the level 1 latent classes, the smallest level 2 class was 1, which contained 17,284 subjects. Finally, the smallest level 1 class was 4, with 11,367 individuals. The confidence that we can place in the latent class memberships produced by this model, based on the entropy value, was reasonably high (recall that values closer to 1 indicate better model fit). In addition, an examination of the average latent class probabilities reveals that for all of the combinations of the level 1 and level 2 latent classes, the mean probabilities were greater than 0.8. Given the complexity of this model and the relatively large number of class combinations (20), these results suggest that the model is producing fairly robust results. These class-specific probabilities appear below.

Average Latent Class Probabilities for Most Likely Latent
Class Pattern (Row)
by Latent Class Pattern (Column)

Latent Class Variable Patterns

Latent Class Pattern No.	CB Class	CW Class
1	1	1
2	1	2
3	1	3
4	1	4
5	1	5
6	2	1
7	2	2
8	2	3
9	2	4
10	2	5
11	3	1
12	3	2
13	3	3
14	3	4
15	3	5
16	4	1
17	4	2
18	4	3
19	4	4
20	4	5

	1	2	3	4	5	6	7	8	9
1	0.865	0.000	0.037	0.033	0.000	0.000	0.000	0.000	0.000
2	0.000	0.781	0.000	0.000	0.114	0.000	0.000	0.000	0.000
3	0.067	0.000	0.820	0.000	0.026	0.000	0.000	0.000	0.000
4	0.044	0.000	0.000	0.919	0.000	0.000	0.000	0.000	0.000
5	0.000	0.015	0.093	0.000	0.795	0.000	0.000	0.000	0.000
6	0.000	0.000	0.000	0.000	0.000	0.801	0.000	0.091	0.012
7	0.000	0.000	0.000	0.000	0.000	0.000	0.920	0.000	0.000
8	0.000	0.000	0.000	0.000	0.000	0.021	0.000	0.810	0.000
9	0.000	0.000	0.000	0.000	0.000	0.064	0.000	0.000	0.828
10	0.000	0.000	0.000	0.000	0.000	0.000	0.063	0.026	0.000
11	0.000	0.000	0.000	0.000	0.000	0.004	0.000	0.001	0.000
12	0.000	0.000	0.000	0.000	0.000	0.000	0.023	0.000	0.000
13	0.000	0.000	0.000	0.000	0.000	0.000	0.000	0.007	0.000
14	0.000	0.000	0.000	0.000	0.000	0.000	0.000	0.000	0.003
15	0.000	0.000	0.000	0.000	0.000	0.000	0.002	0.000	0.000
16	0.023	0.000	0.001	0.002	0.000	0.000	0.000	0.000	0.000
17	0.000	0.004	0.000	0.000	0.001	0.000	0.000	0.000	0.000
18	0.002	0.000	0.013	0.000	0.000	0.000	0.000	0.000	0.000
19	0.001	0.000	0.000	0.032	0.000	0.000	0.000	0.000	0.000
20	0.000	0.000	0.002	0.000	0.007	0.000	0.000	0.000	0.000

	10	11	12	13	14	15	16	17	18
1	0.000	0.000	0.000	0.000	0.000	0.000	0.057	0.000	0.006
2	0.000	0.000	0.000	0.000	0.000	0.000	0.000	0.097	0.000
3	0.000	0.000	0.000	0.000	0.000	0.000	0.002	0.000	0.079
4	0.000	0.000	0.000	0.000	0.000	0.000	0.005	0.000	0.000
5	0.000	0.000	0.000	0.000	0.000	0.000	0.000	0.003	0.004
6	0.000	0.090	0.000	0.003	0.004	0.000	0.000	0.000	0.000
7	0.053	0.000	0.023	0.000	0.000	0.004	0.000	0.000	0.000
8	0.097	0.005	0.000	0.066	0.000	0.002	0.000	0.000	0.000
9	0.000	0.009	0.000	0.000	0.098	0.000	0.000	0.000	0.000
10	0.864	0.000	0.001	0.004	0.000	0.043	0.000	0.000	0.000
11	0.000	0.826	0.000	0.076	0.014	0.000	0.074	0.000	0.003
12	0.001	0.000	0.863	0.000	0.000	0.081	0.000	0.027	0.000
13	0.002	0.032	0.000	0.831	0.000	0.061	0.005	0.000	0.060
14	0.000	0.085	0.000	0.000	0.817	0.000	0.005	0.000	0.000
15	0.014	0.000	0.034	0.052	0.000	0.850	0.000	0.001	0.005
16	0.000	0.034	0.000	0.005	0.000	0.000	0.856	0.000	0.059
17	0.000	0.000	0.081	0.000	0.000	0.005	0.000	0.806	0.000
18	0.000	0.001	0.000	0.049	0.000	0.006	0.046	0.000	0.838
19	0.000	0.004	0.000	0.000	0.024	0.000	0.066	0.000	0.000
20	0.000	0.000	0.003	0.002	0.000	0.065	0.000	0.024	0.070

	19	20
1	0.001	0.000
2	0.000	0.008
3	0.000	0.006
4	0.033	0.000
5	0.000	0.090
6	0.000	0.000
7	0.000	0.000
8	0.000	0.000
9	0.000	0.000
10	0.000	0.000
11	0.002	0.000
12	0.000	0.005
13	0.000	0.002
14	0.090	0.000
15	0.000	0.041
16	0.020	0.000
17	0.000	0.103
18	0.000	0.046
19	0.873	0.000
20	0.000	0.826

Now that we have a sense for the number and size of the various latent classes, we can examine the class-specific means for each, which appears

below. These results have been taken from analyses using the exported file
resultscbcw5.txt and have been placed in the table for ease of interpretation. These additional analyses on the saved data file can be conducted using
any statistical software to the researcher's liking, including SPSS, SAS, and R.

These results are very dense and beyond the scope of what we would go
over in great detail here. However, we can get a sense for how these values
are interpreted by examining the first two combinations of latent classes
(1 1 and 1 2). Here, we see that the mean achievement scores for all three tests
are lower for the 1 1 group than for those in 1 2 group. In addition, those
in combination 1 1 have a lower mean SES index than those in combination 1 2. Thus, we would conclude that combination 1 2 represents relatively
higher-achieving, wealthier students, as compared with combination 1 1.
The researcher would need to go through similar characterizations of the
results in order to complete the full picture of the latent classes contained
herein (Table 10.2).

Finally, we specified a direct relationship between level 2 latent class membership and level 1 latent class membership. The resulting parameter estimates and hypothesis tests assess the extent to which the class to which one's
school belongs is related to the class to which individual students belong.
The Mplus output appears below.

```
Between Level

Categorical Latent Variables

Within Level

    Intercepts
      CW#1                     0.319        0.121        2.630        0.009
      CW#2                    -1.766        0.063      -27.823        0.000
      CW#3                     0.613        0.060       10.143        0.000
      CW#4                    -1.050        0.174       -6.041        0.000

Between Level

  CW#1         ON
    CB#1                       1.993        0.100       19.867        0.000
    CB#2                      -3.916        0.159      -24.681        0.000
    CB#3                      -1.571        0.040      -39.589        0.000

  CW#2         ON
    CB#1                      -0.557        0.191       -2.917        0.004
    CB#2                       1.962        0.059       33.466        0.000
    CB#3                       0.717        0.043       16.757        0.000

  CW#3         ON
    CB#1                       0.960        0.075       12.763        0.000
```

CB#2		-2.192	0.064	-34.300	0.000
CB#3		-0.833	0.025	-32.745	0.000
CW#4	ON				
CB#1		3.108	0.133	23.301	0.000
CB#2		-4.918	0.324	-15.201	0.000
CB#3		-2.280	0.076	-30.079	0.000
Means					
CB#1		-1.167	0.105	-11.158	0.000
CB#2		-1.222	0.144	-8.498	0.000
CB#3		-0.068	0.121	-0.562	0.574

TABLE 10.2

Mean (Standard Deviation) SES Index, Math, Reading, and Science Scores by
Level 1 and Level 2 Latent Classes

Level 2 (School) Latent Class	Level 1 (Student) Latent Class					
	Class 1	Class 2	Class 3	Class 4	Class 5	Overall
Overall						
SES	−0.56 (0.85)	−0.29 (0.83)	0.02 (0.84)	0.38 (0.81)	0.73 (0.76)	0.12 (0.90)
Math	336 (41)	416 (34)	487 (32)	558 (33)	636 (37)	502 (90)
Read	322 (46)	414 (37)	491 (35)	563 (34)	636 (37)	502 (94)
Science	328 (42)	419 (31)	498 (29)	574 (29)	653 (34)	511 (96)
Class 1						
SES	−0.46 (0.73)	−0.38 (0.76)	−0.22 (0.76)	0.04 (0.75)	0.28 (0.71)	−0.05 (0.78)
Math	342 (50)	426 (33)	493 (31)	562 (33)	646 (40)	532 (90)
Read	302 (53)	406 (38)	485 (34)	559 (34)	632 (39)	522 (96)
Science	322 (50)	423 (31)	503 (29)	578 (27)	655 (35)	541 (95)
Class 2						
SES	−0.26 (0.83)	−0.01 (0.79)	0.26 (0.77)	0.54 (0.75)	0.87 (0.72)	0.40 (0.82)
Math	338 (42)	422 (35)	492 (34)	564 (34)	642 (38)	523 (87)
Read	316 (45)	409 (36)	488 (35)	563 (34)	639 (38)	518 (91)
Science	324 (42)	418 (31)	494 (31)	571 (30)	653 (37)	527 (91)
Class 3						
SES	−0.67 (87)	−0.38 (0.80)	−0.04 (0.78)	0.33 (0.77)	0.72 (0.76)	0.07 (0.89)
Math	338 (44)	420 (36)	492 (34)	566 (32)	642 (36)	508 (95)
Read	317 (50)	410 (37)	486 (36)	557 (35)	624 (37)	497 (97)
Science	324 (45)	419 (32)	498 (28)	571 (29)	648 (34)	510 (99)
Class 4						
SES	−0.55 (0.86)	−0.27 (0.85)	0.07 (0.86)	0.48 (0.82)	0.94 (0.70)	0.14 (0.93)
Math	334 (38)	413 (34)	484 (31)	552 (32)	627 (34)	488 (86)
Read	328 (42)	416 (36)	494 (34)	568 (33)	644 (35)	497 (93)
Science	331 (38)	418 (30)	497 (29)	574 (29)	656 (32)	502 (93)

We see a series of estimates from a multinomial logistic regression model, associating membership in each of the school level (CB) latent classes with each of the student level (CW) classes. The reference categories for each variables are the last ones (5 for CW and 4 for CB). As an example of interpreting these results, we can see that individuals attending schools in level 2 latent class 1 are more likely to belong to level 1 latent class 1 than the individuals attending schools in level 2 latent class 4 ($b = 1.993$, $p < 0.001$). Conversely, individuals whose schools are in level 2 latent classes 2 or 3 are less likely to be in level 1 latent class 1 than the individuals attending schools in level 2 latent class 4.

It is clear that interpretation of the results from an MLCA model can be extremely complex and challenging. However, these models can also yield a wealth of information regarding the structure underlying a set of data, particularly with regard to latent groups that might be present at both levels of analysis. For this reason, we feel that they are extremely promising for research practice but must be interpreted with care in order to fully understand the latent class structure at all levels.

Summary

In this chapter, we have extended the modeling of latent variables to the multilevel context. In many instances, these models resembled the ones that we used for observed variables, such as regression (Chapter 3). In other cases, however, we are working with completely new ways of thinking about and modeling our data. In particular, MLCA presents us with some very interesting, potentially informative, and definitely challenging models. Just as with single-level latent variable modeling, the linking of theory to the model and data that we use is of paramount importance. Given the complexity of the methods discussed in this chapter, it can become very easy for the researcher to drift away from the solid footing of established theory into uncharted and deep waters with the help of difficult-to-fit-and-understand models. In this chapter, our goal was only to introduce the interested reader to the many options available in the latent variable modeling context. The full breadth of such topics is well beyond the purview of a single chapter and indeed could easily be contained within an entire book. However, we do hope that this work provides researchers with a starting point from which they can embark on their own exploration of the finer details of multilevel models and methods for latent variables that are of particular interest to them.

11

Bayesian Multilevel Modeling

Bayesian statistical modeling represents a fundamental shift from the frequentist methods of model parameter estimation that we have been using heretofore. This paradigm shift is evident in part through the methodology used to obtain the estimates: Markov Chain Monte Carlo (MCMC) most commonly for the Bayesian approach and maximum likelihood (ML) and restricted maximum likelihood (REML) in the frequentist case. In addition, Bayesian estimation involves the use of prior distributional information that is not present in frequentist-based approaches. Perhaps even more than the obvious methodological differences, however, the Bayesian analytic framework involves a very different view from that traditionally espoused in the likelihood-based literature as to the nature of population parameters. In particular, frequentist-based methods estimate the population parameter using a single value that is obtained using the sample data only. In contrast, in the Bayesian paradigm the population parameter is estimated as a distribution of values, rather than a single number. Furthermore, this estimation is carried out using both the sample data, as well as prior distribution information provided by the researcher. Bayesian methods combine this prior information regarding the nature of the parameter distribution with information taken from the sample data in order to estimate a posterior distribution of the parameter distribution. In practice, when a single value estimate of a model parameter is desired, such as a regression coefficient linking dependent and independent variables, the mean, median, or mode of the posterior distribution is calculated. In addition, standard deviations and density intervals for model parameters can also be estimated from this posterior distribution.

A key component of conducting Bayesian analysis is the specification of a prior distribution for each of the model parameters. These priors can be either one of two types. Informative priors are typically drawn from prior research and will be fairly specific in terms of both their mean and variance. For example, a researcher may find a number of studies in which a vocabulary test score has been used to predict reading achievement. Perhaps across these studies the regression coefficient is consistently around 0.5. The researcher may then set the prior for this coefficient as the normal distribution with a mean of 0.5 and a variance of 0.1. In doing so, (s)he is stating upfront that the coefficient linking these two variables in their own study is likely to be near this value. Of course, such may not be the case, and because the data is also used to obtain the posterior distribution, the prior plays only a partial role in its determination. In contrast to informative priors, noninformative (sometimes referred to as diffuse) priors are not based on prior research. Rather, noninformative priors are deliberately selected so as to constrain the posterior distribution for the parameter as little as possible, in light of the fact that little or no useful information is available for setting the prior distribution. As an example, if there is not sufficient evidence in the literature for the researcher to know what the distribution of the regression coefficient is likely to be, (s)he may set the prior as a normal with a mean of 0 and a large variance of, perhaps, 1000 or even more. By using such a large variance for the prior distribution, the researcher is acknowledging the lack of credible information regarding what the posterior distribution might be, thereby leaving the posterior distribution largely unaffected by the prior, and relying primarily on the observed data to obtain the parameter estimate.

The reader may rightly question why, or in what situations, Bayesian multilevel modeling might be particularly useful, or even preferable to frequentist methods. One primary advantage of Bayesian methods in some situations, including with multilevel modeling, is that unlike ML and REML, it does not rely on any distributional assumptions about the data. Thus, the determination of Bayesian credibility intervals (corresponding to confidence intervals) can be made without worry even if the data come from a skewed distribution. In contrast, ML or REML confidence intervals may not be accurate if foundational distributional assumptions are not met. In addition, the Bayesian approach can be quite useful when the model to be estimated is very complex and frequentist-based approaches such as ML and REML are not able to converge. A related advantage is that the Bayesian approach may be better able to provide accurate model parameter estimates in the small sample case. And, of course, the Bayesian approach to parameter estimation can be used in cases where ML and REML also work well, and as we will see below, the different methods generally yield similar results in such situations.

The scope of this book does not encompass the technical aspects of MCMC estimation, which is most commonly used to obtain Bayesian estimates. The interested reader is encouraged to reference any of several good works on the topic. In particular, Lynch (2010) provides a very thorough introduction to Bayesian methods for social scientists including a discussion of the MCMC algorithm, and Kruschke (2011) provides a nice general description of applied Bayesian analysis. It should be noted here that while MCMC is the most frequently used approach for parameter estimation in the Bayesian context, it is not itself inherently Bayesian. Rather, it is simply an algorithmic approach to sampling from complex sampling distributions, such as a posterior distribution that might be seen with complicated models such as those in the multilevel context. Although we will not describe the MCMC process in much detail here, it is necessary to discuss conceptually how the methodology works, in part so that you, the reader, might be more comfortable with where the parameter estimates come from, and in part because we will need to diagnose whether the method has worked appropriately so that we can have confidence in the final parameter estimates.

MCMC is an iterative process in which the prior distribution is combined with information from the actual sample in order to obtain an estimate of the posterior distributions for each of the model parameters (e.g., regression coefficients and random effect variances). From this posterior distribution, parameter values are simulated a large number of times in order to obtain an estimated posterior distribution. After each such sample is drawn, the posterior is updated. This iterative sampling and updating process is repeated a very large number of times (e.g., 10,000 or more) until there is evidence of convergence regarding the posterior distribution; that is, a value from one sampling draw is very similar to the previous sample draw. The Markov Chain part of MCMC reflects the process of sampling a current value from the posterior distribution, given the previous sampled value, while Monte Carlo reflects the random simulation of these values from the posterior distribution. When the chain of values has converged, we are left with an estimate of the posterior distribution of the parameter of interest (e.g., regression coefficient). At this point, a single model parameter estimate can be obtained by calculating the mean, median, or mode from the posterior distribution.

When using MCMC, the researcher must be aware of some technical aspects of the estimation that need to be assessed in order to ensure that the analysis has worked properly. The collection of 10,000 (or more) individual parameter estimates form a lengthy time series, which must be examined to ensure that two things are true. First, the parameter estimates must converge and second the autocorrelation between different iterations in the process should be low. Parameter convergence can be assessed through the use of a trace plot, which is simply a graph of the parameter estimates in order from the first iteration to the last. The autocorrelation of estimates is calculated for a variety of iterations, and the researcher will look for the distance between estimates at which the autocorrelation becomes quite low. When it is determined at what point the autocorrelation between estimates is sufficiently low, the estimates are thinned so as to remove those that might be more highly autocorrelated with one another than would be desirable. So, for example, if the autocorrelation is low when the estimates are 10 iterations apart, the time series of 10,000 sample points would be thinned to include only every 10th observation, in order to create the posterior distribution of the parameter. The mean/median/mode of this distribution would then be calculated using only the thinned values, in order to obtain the single parameter estimate value that is reported by R. A final issue in this regard is what is known as the burn in period. Thinking back to the issue of distributional convergence, the researcher will not want to include any values in the posterior distribution for iterations prior to the point at which the time series converged. Thus, iterations prior to this convergence are referred to as having occurred during the burn in period, and are not used in the calculation of posterior means/medians/modes. Each of these MCMC conditions (number of iterations, thinning rate, and burn in period) can be set by the user in R, or default values can be used. In the remainder of this chapter we will provide detailed examples of the diagnosis of MCMC results, and the setting not only of MCMC parameters, but also of prior distributions.

Mplus Multilevel Modeling for a Normally Distributed Response Variable

We will begin our discussion of fitting a random intercept model with the Prime Time data, which we have used in numerous examples in previous chapters. In particular, we will fit a model in which reading achievement score is the dependent variable, and vocabulary score is the independent variable. Students are nested within schools, which we will treat as a random effect. Much of the programming for Bayesian multilevel modeling in Mplus will look very similar to the standard multilevel modeling that we have seen in earlier chapters. As we discussed previously in this chapter, a key component of Bayesian modeling is the use of prior distribution information in the estimation of the posterior distribution of the model parameters. Mplus has a default set of priors that it uses for each model parameter, and upon which we will rely for the first example analysis. The default priors for the model coefficients and intercepts are noninformative in nature, taken from the standard normal distribution with a mean of 0, and a variance of infinity. This very large variance for the prior reflects our relative lack of confidence that the mean of the coefficient distributions is in fact 0. The prior distributions for the random effects are Inverse Gamma, or IG (−1,0).

In order to fit the random intercept model with a single predictor under the Bayesian framework with default priors, we will use the following Mplus program:

```
TITLE:      Model 10.1  Two-level regression model
        with random intercept
DATA:FILE = prime_time_10.csv;
VARIABLE:  NAMES = school
geread
gevocab
npamem
senroll;
    usevariables school geread gevocab;
    WITHIN = gevocab;
    CLUSTER = school;
DEFINE:    CENTER gevocab (GRANDMEAN);
ANALYSIS:  TYPE = TWOLEVEL;
            estimator=bayes;
            chains=2;
            biterations=100000;
```

```
MODEL:
      %WITHIN%
      geread ON gevocab;

output:      tech1 tech8 standardized;

plot:    type=plot2;
```

As noted previously, the bulk of this program is identical to what we would use when fitting a multilevel model with the REML estimator. Therefore, we will focus only on the portions that are unique to the Bayesian context. In particular, the following commands specify how we would like for the model to be fit.

```
estimator=bayes;
chains=2;
biterations=100000;
```

We first must specify that we want to use the Bayesian estimator, followed by a request for two parallel MCMC chains to be created. This is the default used by Mplus, but we include the subcommand here so that the reader can see where to change this value, if they would like to do so. Finally, we request that Mplus allow for a *maximum* of 100,000 MCMC iterations. Mplus uses the potential scale reduction (PSR) factor to automatically determine if convergence has occurred. Using the parallel chains, this criterion involves comparing the within chain variance to the between chain variance for each model parameter. The statistic that reflects this criterion is PSR:

$$PSR = \sqrt{\frac{W_p + B_p}{B_p}} \qquad (11.1)$$

where:

W_p is within chain variance for parameter p
B_p between chain variance for parameter p

The model is assumed to have converged when the PSR is very close to 1. By default, Mplus uses the PSR to determine if and when the MCMC algorithm has converged. It does this by monitoring the largest PSR value from across the parameters at each iteration. When the largest value is below the 1.05 threshold, then the model is said to have converged. In other words, when all of the PSRs from across the model parameters are below 1.05, then Mplus stops the MCMC algorithm and assumes the model has converged. It is also possible for us to request the PSR values be included in the output by requesting tech8 in the output command. Finally, by specifying plot2 in the plot command, we will have access to a variety of graphical tools that will prove helpful in evaluating our model. By default, Mplus uses the first half of the iterations to serve as the burn in period and then takes the second half to create the posterior distribution. In addition, the default setting in Mplus is to not thin the MCMC draws to create the posterior. The user can control the thinning using something like thin=10, if she would like to use every 10th MCMC draw in the post-burn in data to create the posterior distribution. The thin subcommand is placed under the analysis command, as is estimator, chains, and biterations.

When interpreting the results of the Bayesian analysis, we first want to know whether we can be confident in the quality of the parameter estimates for both the fixed and random effects. We can access these plots after we run the analysis and requesting plot2, by clicking on the 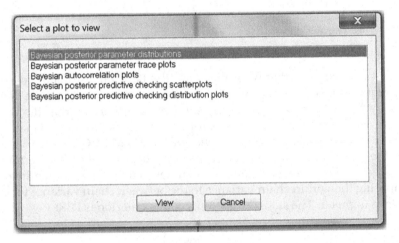 menu button. We are then presented with the following window, from which we can select the graphs that are of interest.

Select a plot to view

Bayesian posterior parameter distributions
Bayesian posterior parameter trace plots
Bayesian autocorrelation plots
Bayesian posterior predictive checking scatterplots
Bayesian posterior predictive checking distribution plots

View   Cancel

If we would like to examine the trace plot for one or more parameters to ensure that convergence was attained, we would simply click on Bayesian posterior parameter trace plots, click View, and then select the parameter of interest from the pull down bar in the window below.

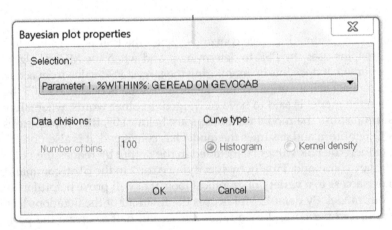

Bayesian plot properties

Selection:

Parameter 1. %WITHIN%: GEREAD ON GEVOCAB

Data divisions: Curve type:

Number of bins   100   ◉ Histogram   ○ Kernel density

OK   Cancel

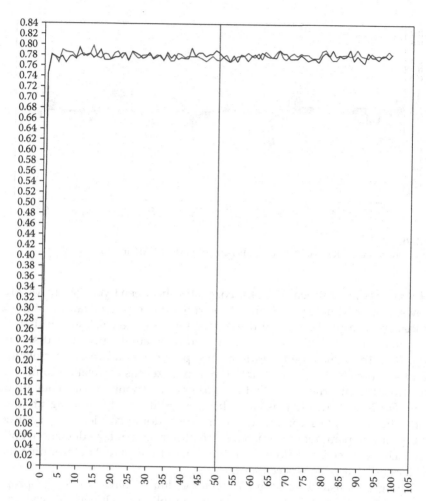

FIGURE 11.1
Trace plot for fixed effects coefficient linking GEREAD and GEVOCAB.

The trace plot for the parameter linking geread and gevocab appears below. The x-axis displays the MCMC iteration, the y-axis displays the parameter estimate, and the two chains are represented by the different colored lines in the graph. It is clear from this plot that convergence has been reached, as both curves are centered on the same value along the y-axis (Figure 11.1).

Using the same steps as we did to obtain the trace plot, we can also examine the autocorrelation function for each model parameter. The autocorrelation is simply the correlation coefficient for MCMC sample values for a given parameter that are a certain number of steps apart. As an example, consider Figure 11.2, which contains the autocorrelation function plot of the fixed effect coefficient estimate for the first chain. On the x-axis are the lags, or the number of steps between MCMC estimates, and on the y-axis is the correlation.

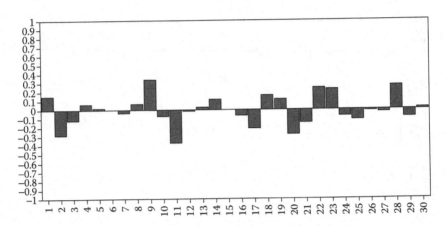

FIGURE 11.2
Autocorrelation plot for fixed effects coefficient linking GEREAD and GEVOCAB.

From this plot, it appears that the correlation between two adjacent MCMC draws is approximately 0.15, and the correlation between draws that are 2 spaces apart (e.g., draw 1 and draw 3) is approximately 0.3. Recall that by default, Mplus does not thin the data used to create the posterior distribution. Thus, the values used to estimate this parameter had an autocorrelation of approximately 0.15. If we wanted to minimize this correlation to near 0, we might thin the data such that the posterior distribution was made up of every 5th MCMC draw. This could be done easily in Mplus using `thin=5` under the `analysis` command. In this case, doing so yielded very little change in the parameter estimate, or the posterior standard deviation, both of which are discussed below. Therefore, it would seem that not thinning the posterior distribution is acceptable.

Having established that the parameter estimates have converged properly, and that our rate of thinning in the sampling of MCMC derived values yields acceptable estimates, we are now ready to examine the specific parameter estimates for our model. The relevant output for this analysis appears below.

```
MODEL FIT INFORMATION

Number of Free Parameters                              4
Bayesian Posterior Predictive Checking using Chi-Square

                95% Confidence Interval for the Difference Between
                the Observed and the Replicated Chi-Square Values

                        -14.057                5.439

        Posterior Predictive P-Value               0.583
```

```
Information Criteria

            Deviance (DIC)                           105228.785
            Estimated Number of Parameters (pD)        122.394

STANDARDIZED MODEL RESULTS

STDYX Standardization
```

	Estimate	S.D.	Posterior One-Tailed P-Value	95% C.I. Lower 2.5%	Upper 2.5%	Significance
Within Level						
GEREAD ON						
GEVOCAB	0.768	0.003	0.000	0.762	0.774	*
Residual Variances						
GEREAD	0.410	0.005	0.000	0.401	0.419	*
Between Level						
Means						
GEREAD	1.101	0.130	0.000	0.896	1.388	*
Variances						
GEREAD	1.000	0.000	0.000	1.000	1.000	

We are first given information about model fit, including the deviance information criterion (DIC), which can be used for comparing various models and selecting the one that provides optimal fit, and the posterior predictive *p*-value (PPP). The DIC is interpreted in much the same fashion as the Akaike information criterion (AIC) and Bayesian information criterion (BIC), which we discussed in earlier chapters, and for which smaller values indicate better model fit. The PPP serves as a measure of the goodness of fit of the model parameter estimates to the data. In order to calculate the PPP, the item parameters that were estimated by the model are used to simulate a large number (e.g., 1000) of datasets of the same size as the actual data. For each of these simulated datasets, the Chi-square goodness of fit test is calculated, creating a distribution of this statistic across the 1000 simulated datasets. This distribution represents what would be expected if the null hypothesis of good model fit holds. The Chi-square goodness of fit statistic for the actual data is then compared to the distribution of PPP values for the generated data. The PPP value itself represents the probability that a sample drawn from the population would produce a test statistic that exceeds that of the current sample. Therefore, values at the extremes (near 0 or 1) are indicative of poor model fit to the data. Generally speaking, values closer to 0.5 are indicative of optimal model fit, as they suggest that half of the samples in the population should yield Chi-square values greater than that of the current sample, and half should yield smaller values. The extent to which the PPP can be seen as deviating from this 0.5 standard is a matter of subjective judgment by the researcher (Kaplan 2014). In addition to the PPP, Mplus also provides a 95% credibility interval for the difference between the observed Chi-square test statistic, and those for the simulated data. If this interval includes 0, we may conclude that the model does provide good fit to the data. For this particular example, both the PPP and the confidence interval for Chi-square difference values would indicate that the model does yield good fit to the data.

We are then provided with the mean of the posterior distributions for each of the model effects. We present the standardized results above. The mean variance estimate for the school random effect is 0.410, with a 95% credibility interval of 0.401–0.419. Remember that we interpret credibility intervals in Bayesian modeling in much the same way that we interpret confidence intervals in frequentist modeling. This result indicates that reading achievement scores do differ across schools, because 0 is not in the interval. Similarly, the residual variance also differs from 0. With regard to the fixed effect of vocabulary score, which had a mean posterior value of 0.768, we also conclude that the results are statistically significant, given that 0 is not in its 95% credibility interval (0.762, 0.774). We also have a *p*-value for this effect and the intercept, both of which are significant with values less than 0.05. The positive value of the posterior mean indicates that students with higher vocabulary scores also had higher reading scores.

In order to demonstrate how we can change the number of iterations and the rate of thinning in Mplus, we will re-estimate Model 10.1 with 1000 iterations and a thinning rate of 5. The analysis section of the Mplus program for fitting this model, followed by the relevant output, appears below.

```
ANALYSIS:  TYPE = TWOLEVEL;
           estimator=bayes;
           chains=2;
           thin=5;
           fbiterations=1000;
```

As with the initial model, all parameter estimates appear to have successfully converged. The results, in terms of the posterior means, are also very similar to what we obtained using the default values for the number of iterations, the burn in period, and the thinning rate. This result is not surprising, given that the diagnostic information for our initial model were all very positive. Nonetheless, it was useful for us to see how the default values can be changed if we need to do so.

```
MODEL FIT INFORMATION

Number of Free Parameters                              4

Bayesian Posterior Predictive Checking using Chi-Square

          95% Confidence Interval for the Difference Between
          the Observed and the Replicated Chi-Square Values

                          -8.163                7.468

          Posterior Predictive P-Value          0.510

Information Criteria

          Deviance (DIC)                       105215.227
          Estimated Number of Parameters (pD)    117.969

STANDARDIZED MODEL RESULTS

STDYX Standardization

                         Posterior  One-Tailed        95% C.I.
              Estimate     S.D.      P-Value   Lower 2.5%  Upper 2.5%  Significance
Within
Level

 GEREAD  ON
   GEVOCAB    0.768       0.003      0.000      0.761       0.774           *

 Residual
 Variances
   GEREAD     0.410       0.005      0.000      0.400       0.421           *

Between
Level

 Means
   GEREAD     1.104       0.140      0.000      0.850       1.396           *

 Variances
   GEREAD     1.000       0.000      0.000      1.000       1.000
```

Including Level 2 Predictors with MCMCglmm

In addition to understanding the extent to which reading achievement is related to vocabulary test score, in Chapter 3 we were also interested in the relationship of school (senroll), a level 2 variable, and reading achievement. Including a level 2 variable in the analysis with Mplus is very straightforward, as can be seen in the following program.

```
TITLE:     Model 10.3  Two-level regression model
        with random intercept and a level 2
        predictor
DATA: FILE = prime_time_10.csv;
VARIABLE:    NAMES = school
geread
gevocab
npamem
senroll;
    usevariables school geread gevocab senroll;
      WITHIN = gevocab;
    BETWEEN = senroll;
     CLUSTER = school;
DEFINE:    CENTER gevocab (GRANDMEAN);
ANALYSIS: TYPE = TWOLEVEL;
            estimator=bayes;
            chains=2;

MODEL:
        %WITHIN%
        geread ON gevocab;

    %BETWEEN%
    geread ON senroll;

output:      tech1 tech8 standardized;

plot:    type=plot2;
```

An examination of the trace plots shows that we achieved convergence for all parameter estimates. Figure 11.3 contains the trace plot for the level 2 variable, senroll.

The autocorrelations appear below the graphs and reveal that the default thinning rate (none) may not be sufficient so as to remove autocorrelation from the estimates for school enrollment. It would appear that a thinning rate of 2 will be sufficient, however (Figure 11.4).

Given the elevated autocorrelation for a lag of 1, we refit the model with 10,000 iterations and a thinning rate of 2 (Model 10.4).

The summary results for the model with 1000 iterations and a thinning rate of 2 appear below. It should be noted that the trace plots and histograms of parameter estimates for this model indicated that convergence had been attained. From these results we can see that overall fit, based on the DIC, is very similar to that of the model not including senroll. The posterior mean estimate and associated 95% credible interval for this parameter show that senroll was statistically significantly related to reading achievement, and was negative suggesting that attending a school with larger enrollment was associated with lower reading achievement.

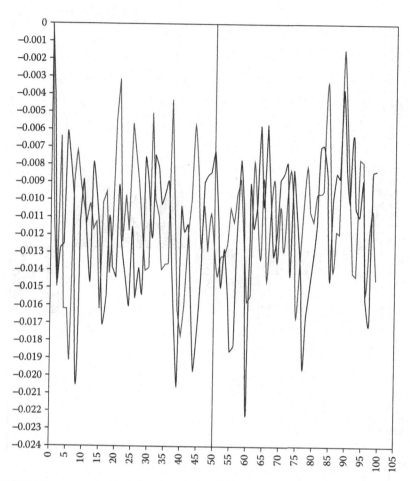

FIGURE 11.3
Trace plot for level 2 effect coefficient linking SENROLL and GEVOCAB.

```
MODEL FIT INFORMATION

Number of Free Parameters                          5

Bayesian Posterior Predictive Checking using Chi-Square

        95% Confidence Interval for the Difference Between
        the Observed and the Replicated Chi-Square Values

                        -9.028              9.259

        Posterior Predictive P-Value              0.480
```

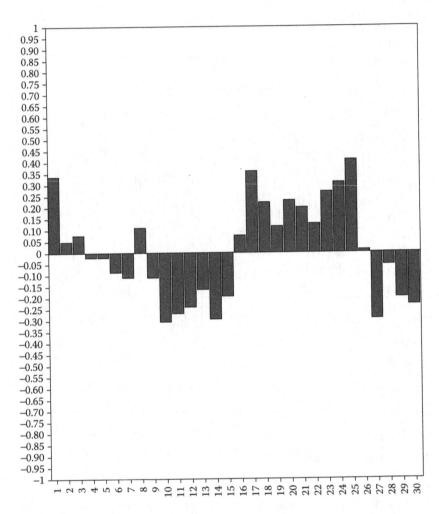

FIGURE 11.4
Autocorrelation plot for level 2 coefficient linking GEREAD to SENROLL.

```
Information Criteria

              Deviance (DIC)                           105220.597
              Estimated Number of Parameters (pD)       121.311

STANDARDIZED MODEL RESULTS

STDYX Standardization
                        Posterior  One-Tailed        95% C.I.
              Estimate   S.D.      P-Value   Lower 2.5%  Upper 2.5%  Significance
Within
Level

  GEREAD  ON
    GEVOCAB    0.768     0.003     0.000      0.761       0.774          *
```

Residual Variances						
GEREAD	0.410	0.005	0.000	0.400	0.420	*
Between Level						
GEREAD ON						
SENROLL	-0.266	0.087	0.000	-0.425	-0.084	*
Intercepts						
GEREAD	1.912	0.301	0.000	1.307	2.505	*
Residual Variances						
GEREAD	0.929	0.047	0.000	0.820	0.993	*

As a final separate example in this section, we will fit a random coefficients model, in which we allow the relationship of vocabulary score and reading achievement to vary across schools. The Mplus code for fitting this model appears below. Note that when the random option is used with the bayes estimator, standardized parameter estimates cannot be obtained.

```
TITLE:     Model 10.5  Two-level regression model
          with random intercept and random slope
DATA:  FILE = prime_time_10.csv;
VARIABLE:   NAMES = school
geread
gevocab
npamem
senroll;
    usevariables school geread gevocab;
     WITHIN = gevocab;
     CLUSTER = school;
DEFINE:    CENTER gevocab (GRANDMEAN);
ANALYSIS: TYPE = TWOLEVEL random;
          estimator=bayes;
          chains=2;

MODEL:
     %WITHIN%
   s | geread ON gevocab;

   %BETWEEN%
   geread with s;

output:     tech1 tech8;

plot:   type=plot2;
```

The trace and autocorrelation plots (not shown here) indicate that the parameter estimation has converged properly, and that the thinning rate appears to be satisfactory for removing autocorrelation from the estimate values. The model results appear below. First, we should note that the DIC for this random coefficients model is smaller than that of the random intercepts only models above. The estimate of the random coefficient for vocabulary is 0.258, with a 95% credible interval of −0.220 to 0.648. Because this interval includes 0, we can conclude that the random coefficient is not different from 0 in the population, and that the relationship between reading achievement and vocabulary test score does not vary from one school to another.

```
MODEL FIT INFORMATION

Number of Free Parameters                              6

Information Criteria

        Deviance (DIC)                         100700.487
        Estimated Number of Parameters (pD)       161.476
```

```
MODEL RESULTS
                    Posterior  One-Tailed        95% C.I.
          Estimate    S.D.    P-Value    Lower 2.5%  Upper 2.5%  Significance

Within
Level

Residual
Variances
  GEREAD   579.468   7.419    0.000       566.769     596.430         *

Between
Level

GEREAD  WITH
  S       -17.363    2.260    0.000       -22.530     -13.326         *

Means
  GEREAD     6.250   0.609    0.000         5.159       7.619         *
  S          0.258   0.231    0.150        -0.220       0.648

Variances
  GEREAD    45.970   7.388    0.000        34.122      63.946         *
  S          6.650   1.031    0.000         4.924       8.819         *
```

User Defined Priors

Finally, we need to consider the situation in which the user would like to provide her or his own prior distribution information, rather than rely on the defaults established in Mplus. To do so, we will make use of the `model` priors command. Consider the situation where the researcher has informative priors for one of the model parameters. As an example let us assume that a number of studies in the literature report finding a small but consistent positive relationship between reading achievement and vocabulary. In order to incorporate this informative prior into a model relating these two variables, we would first need to define our prior, as below. This can be done by labeling the parameter of interest and then defining the priors for that parameter using the model priors command. In this case, we set the prior mean and variance of the coefficient for memory to 0.5 and 0.1, respectively. We select a fairly small variance for the working memory coefficient because we have much prior evidence in the literature regarding the anticipated magnitude of this relationship.

```
TITLE:      Model 10.1_informative  Two-level regression model
            with random intercept and informative priors for the
            model coefficient
DATA: FILE = prime_time_10.csv;
VARIABLE:  NAMES = school
geread
gevocab
npamem
senroll;
     usevariables school geread gevocab;
      WITHIN = gevocab;
      CLUSTER = school;
DEFINE:     CENTER gevocab (GRANDMEAN);
ANALYSIS:   TYPE = TWOLEVEL;
                estimator=bayes;
                chains=2;

MODEL:

        %WITHIN%
        geread ON gevocab (p1);

model priors:
    p1~N(0.5,0.1);

output:       tech1 tech8 standardized;

plot:    type=plot2;
```

The model appears to have converged well, and the autocorrelations suggest that the rate of thinning was appropriate (Figures 11.5 and 11.6).

The summary of the model fit results appear below. Of particular interest is the coefficient for the fixed effect numsense. The standardized posterior mean is 0.768, with a credible interval ranging from 0.762 to 0.774, indicating that the relationship between working memory and reading achievement is

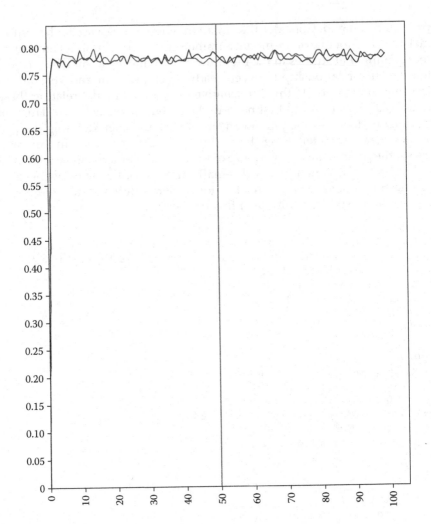

FIGURE 11.5
Trace plot of level 1 coefficient linking GEREAD and GEVOCAB with user defined priors.

statistically significant. In addition, this value is very similar to that obtained with the noninformative priors in Model 10.1. In this case, because the sample is so large, the effect of the prior on the posterior distribution is very small. The impact of the prior would be much greater were we working with a smaller sample.

```
Bayesian Posterior Predictive Checking using Chi-Square

        95% Confidence Interval for the Difference Between
        the Observed and the Replicated Chi-Square Values

                -14.066                    5.416
```

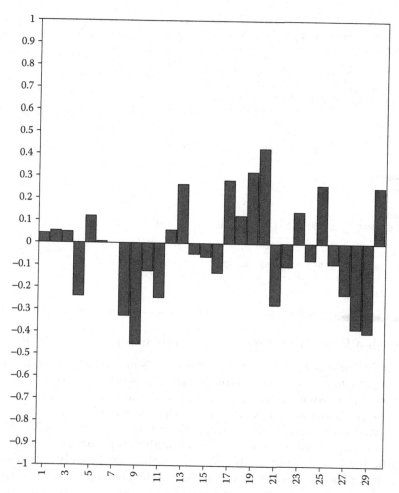

FIGURE 11.6
Autocorrelation plot of level 1 coefficient linking GEREAD and GEVOCAB with user defined priors.

```
        Posterior Predictive P-Value                0.583

Information Criteria
        Deviance (DIC)                         105228.745
        Estimated Number of Parameters (pD)       122.379

STANDARDIZED MODEL RESULTS


STDYX Standardization
```

	Posterior	One-Tailed		95% C.I.		
	Estimate	S.D.	P-Value	Lower 2.5%	Upper 2.5%	Significance
Within Level						
GEREAD ON						
GEVOCAB	0.768	0.003	0.000	0.762	0.774	*
Residual Variances						
GEREAD	0.410	0.005	0.000	0.401	0.419	*
Between Level						
Means						
GEREAD	1.101	0.130	0.000	0.896	1.388	*
Variances						
GEREAD	1.000	0.000	0.000	1.000	1.000	

Bayesian Estimation for a Dichotomous-Dependent Variable

Mplus can also be used to fit multilevel models in which the outcome variable is dichotomous in nature. In most respects, the use of the software will be very similar to what we have seen with a continuous outcome, as described previously. Therefore, we will focus on aspects of model fitting that differ from what we have seen up to this point. Our first example involves fitting a model for a dichotomous-dependent variable using Bayesian multilevel logistic regression. Specifically, the model of interest involves predicting whether or not a student receives a passing score on a state math assessment (score2) as a function of their number sense (numsense) score on a formative math assessment. Following is the Mplus code for fitting this model.

```
TITLE:     Model 10.6  Two-level logistic regression
       model with random intercept
DATA: FILE = mathfinal10.csv;
VARIABLE: NAMES = numsense
score2
school;
    usevariables school numsense score2;
    WITHIN = numsense;
    CLUSTER = school;
    categorical = score2;
    missing are .;
DEFINE:    CENTER numsense (GRANDMEAN);
```

```
ANALYSIS:   TYPE = TWOLEVEL;
            estimator=bayes;
            chains=2;
            thin=4;
            fbiterations=1000;

MODEL:
      %WITHIN%
      score2 ON numsense;

output:    tech1 tech8 standardized;

plot:    type=plot2;
```

From the trace plots and histograms, convergence was somewhat questionable, even though the automated procedure did indicate convergence after 300 iterations. In addition, the autocorrelation plot suggested that thinning of every 4th observation in the MCMC sample would ensure that there was not autocorrelation among the sampled values. Thus, the model was rerun with thin=4 and fbiterations=1000. Following are the trace and autocorrelation plots for the coefficient linking numsense to score2 (Figures 11.7 and 11.8).

The model fit information, in form of PPP, suggests that the fit of the model is reasonable, as does the fact that the 95% confidence interval for the difference in Chi-square values includes 0. In terms of model parameter estimation results, the number sense score was found to be statistically significantly related to whether or not a student received a passing score on the state mathematics assessment. The posterior mean for the coefficient is 0.641, indicating that the higher an individual's number sense score, the greater the likelihood that (s)he will pass the state assessment.

```
MODEL FIT INFORMATION

Number of Free Parameters                           3

Bayesian Posterior Predictive Checking using Chi-Square

          95% Confidence Interval for the Difference Between
          the Observed and the Replicated Chi-Square Values

                    -4.692              6.072

          Posterior Predictive P-Value        0.417

STANDARDIZED MODEL RESULTS


STDYX Standardization
```

		Posterior	One-Tailed	95% C.I.		
		Estimate	S.D.	P-Value	Lower 2.5%	Upper 2.5%	Significance
Within Level							
SCORE2	ON						
NUMSENSE		0.641	0.007	0.000	0.626	0.657	*
Between Level							
Thresholds							
SCORE2$1		-0.666	0.216	0.000	-1.074	-0.304	*
Variances							
SCORE2		1.000	0.000	0.000	1.000	1.000	

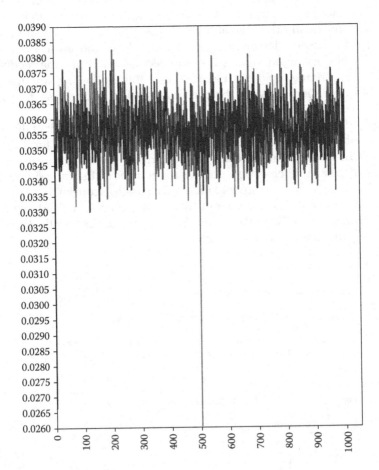

FIGURE 11.7
Trace plot for level 1 effect coefficient linking SCORE2 and NUMSENSE.

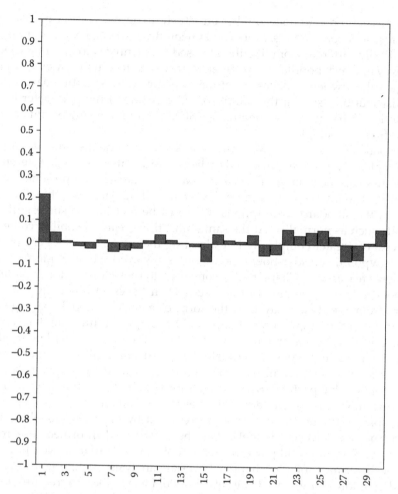

FIGURE 11.8
Autocorrelation plot for level 1 coefficient linking SCORE2 to NUMSENSE.

Essentially the same program that was used here would also be used to fit a model for an ordinal variable with more than two categories.

Summary

The material presented in this chapter represents a marked departure from that presented in the first 9 chapters of the book. In particular, methods presented in the earlier chapters were built upon a foundation of ML estimation. Bayesian modeling, which is the focus of Chapter 10, leaves likelihood-based analyses behind, and instead relies on MCMC to derive a posterior distribution for model parameters. More fundamentally, however, Bayesian statistics is radically different from likelihood-based frequentist statistics in terms of the way in which population parameters are estimated. In the latter they take single values, whereas Bayesian estimation estimates population parameters as distributions, so that the sample-based estimate of the parameter is the posterior distribution obtained using MCMC, and not a single value calculated from the sample.

Beyond the more theoretical differences between the methods described in Chapter 10 and those presented earlier, there are also very real differences in terms of application. Analysts using Bayesian statistics are presented with what is, in many respects, a more flexible modeling paradigm that does not rest on the standard assumptions that must be met for the successful use of ML, such as normality. At the same time, this greater flexibility comes at the cost of greater complexity in estimating the model. From a very practical viewpoint, consider how much more is involved when conducting the analyses featured in Chapter 10 as compared to those featured in Chapter 3. In addition, interpretation of the results from Bayesian modeling requires more from the data analyst, in the form of ensuring model convergence, deciding on the length of the chains, and the degree of thinning that should be done, as well as what summary statistic of the posterior will be used to provide a single parameter estimate. Last, and not at all least, the analyst must consider what the prior distributions of the model parameters should be, knowing that particularly for small samples, the choice will have a direct bearing on the final parameter estimates that are obtained.

In spite of the many complexities presented by Bayesian modeling, it is also true that Bayesian models offer the careful and informed researcher with a very flexible and powerful set of tools. In particular, as we discussed at the beginning of the chapter, Bayesian analysis, including multilevel models, allows for greater flexibility in model form, does not require distributional assumptions, and may be particularly useful for smaller samples. Therefore, we can recommend this approach wholeheartedly, with the realization that the interested researcher will need to invest greater time and energy into deciding on priors, determining when/if a model has converged, and what summary statistic of the posterior distribution is most appropriate. Given this investment, you have the potential for some very useful and flexible models that may work in situations where standard likelihood-based approaches do not. Finally, it should be noted that as of the writing of this text, the Bayesian estimator cannot be used to fit models with count variables or survival models.

Appendix: A Brief Introduction to Mplus

Reading Data into Mplus

Data can be read into Mplus from text files of various types (i.e., .dat, .txt, .csv). If each of the columns, which represent the variables, are separated by spaces, tabs, or commas, Mplus is able to automatically read the file correctly. If, on the other hand, the data are formatted in a special fashion, users can use the FORMAT subcommand under the DATA command to read the file. Mplus uses Fortran language formatting in this case. As an example, if the data file opens with a 5 digit integer ID variable, followed by 5 spaces, and then 10 integer variables, each of which takes up 2 columns, the format statement to read in the data would look as follows:

```
FORMAT IS 1F5.0, 5x, 10F2.0;
```

Missing Data

Missing data must be explicitly specified in Mplus. Unlike is the case with SPSS or SAS, for example, Mplus does not have any default missing data indicators. If the researcher codes missing values for all variables as a., the following MISSING command would appear in the DATA portion of the program.

```
MISSING IS.;
```

On the other hand, if the researcher uses 99 to indicate missing values for variables x1, x2, and x3, and 999 for variable y, the MISSING statement would be written as follows:

```
MISSING ARE X1-X3 (99) Y (999);
```

If missing values are simply blank, we can use the following command to denote this.

```
MISSING = BLANK;
```

Types of Data

Mplus recognizes two types of variables: categorical and continuous. All variables must be numeric. Thus, users will need to be sure that they do not have any character or string variables in their data set. Categorical variables are specified by using the CATEGORICAL subcommand under the VARIABLE command.

Restructuring Data into Person Period Format

The first step in fitting multilevel longitudinal models with Mplus is to make sure the data are in the proper longitudinal structure. Oftentimes such data are entered in what is called person-level data structure. Person-level data structure includes one row for each individual in the data set and one column for each variable or measurement on that individual. In the context of longitudinal data, this would mean that each measurement in time would have a separate column of its own. Although person-level data structure works well in many instances, in order to apply multilevel modeling techniques to longitudinal analyses, the data must be reformatted into what is called person period data structure. In this format, rather than having one row for each individual, person period data has one row for each time that each subject is measured, so that data for an individual in the sample will consist of as many rows as there were measurements made. Unfortunately, Mplus does not provide capability to easily restructure data from person-level format to person period format. This restructure is most easily accomplished in a separate program before the data is read into Mplus for analysis. The following sections in this appendix demonstrate how to accomplish this data restructure using three common software options: SPSS, SAS, and R. Given that it is available for free to all researchers using essentially any computer platform, we will demonstrate the use of R for this purpose.

The data to be used in the following examples is the same data used in Chapter 6. Here, we will be starting from the person-level version of this data set (Lang.csv) and restructuring it into the file used in Chapter 6 (LanguagePP.csv). Recall that this file includes the total language achievement test score measured over six different measurement occasions (the outcome variable), four language subtest scores (writing process and features, writing applications, grammar, and mechanics), and variables indicating student ID and school ID. In this file, ID, school, process, application, grammar, and mechanics are all time invariant variables. The language scores measured over the six measurement occasions are time varying variables.

Restructuring into Person Period Format Using R

Restructuring person-level data into person period format in R can be accomplished by creating a new data frame from the person-level data using the stack command. All time invariant variables will need to be copied into the new data file, while time variant variables (e.g., all test scores measured over the six measurement occasions) will need to be stacked in order to create person period format. The following R command will rearrange the data into the necessary format.

```
LanguagePP <- data.frame(ID=Lang$ID, school=Lang$school,
Process=Lang$Process,
Application=Lang$Application, Grammar=Lang$Grammar,
Mechanics=Lang$Mechanics,
stack(Lang, select=LangScore1:LangScore6))
```

This code takes all of the time invariant variables directly from the raw person-level data, while also consolidating the repeated measurements into a single variable called values, and also creates a variable measuring time called ind. At this point we may wish to do some recoding and renaming of variables. Renaming of variables can be accomplished via the names function, and recoding can be done via recode(var, recodes, as.factor.result, as.numeric.result=TRUE, levels). For instance, we could rename the values variable to Language. The values variable is the 7th column, so in order to rename it we would use the following R code:

```
names(LangPP)[c(7)] <- c("Language")
```

We may also wish to recode the dedicated time variable, ind. Currently, this variable is not recorded numerically, but takes on the values "LangScore1," "LangScore2," "LangScore3," "LangScore4," "LangScore5," and "LangScore6." Thus we may wish to recode the values to make a continuous numeric time predictor, as follows:

```
LangPP$Time <- recode(LangPP$ind,
  '"LangScore1"=0; "LangScore2"=1; "LangScore3"=2;
"LangScore4"=3; "LangScore5"=4; "LangScore6"=5;', as.factor.
result=FALSE)
```

The option as.factor.result=FALSE tells R that the resulting values should be considered continuous. Thus, we have not only recoded the ind variable into a continuous time predictor, but also renamed it from ind to Time, and rescaled the variable such that the first time point is 0. As we noted earlier, when time is rescaled in this manner the intercept can be interpreted as the predicted outcome for baseline or time zero.

References

Agresti, A. (2002). *Categorical Data Analysis*. Hoboken, NJ: John Wiley & Sons.

Asparouhov, T. (2014). Continuous time survival analysis in Mplus. Technical report, retrieved from https://www.statmodel.com/download/Survival.pdf.

Asparouhov, T. and Muthén, B. (2008). Multilevel mixture models. In G.R. Hancock and K.M. Samuelsen (Eds.), *Advances in Latent Variable Mixture Models*, pp. 27–51. Charlotte, NC: Information Age Publishing, Inc.

Bickel, R. (2007). *Multilevel Analysis for Applied Research: It's Just Regression!* New York: The Guilford Press.

Bollen, K.A. (1989). *Structural Equations with Latent Variables*. New York: Jon Wiley & Sons.

Brown, T.A. (2015). *Confirmatory Factor Analysis for Applied Research*. 2nd ed., New York: The Guilford Press.

de Leeuw, J. and Meijer, E. (2008). *Handbook of Multilevel Analysis*. New York: Springer.

Finch, W.H. (2011). A Comparison of Factor Rotation Methods for Dichotomous Data. *Journal of Modern Applied Statistical Methods, 10*(2), 549–570.

Finch, W.H. and French, B.F. (2014). Multilevel latent class analysis: Parametric and nonparametric models. *Journal of Experimental Education, 82*(3), 307–333.

Finney, S.J. and DiStefano, C. (2013). Nonnormal and Categorical Data in Structural Equation Models. In G.R. Hancock and R.O. Mueller (Eds.), *A Second Course in Structural Equation Modeling*, 2nd ed., pp. 439–492. Charlotte, NC: Information Age.

Flora, D.B. and Curran, P.J. (2004). An empirical evaluation of alternative methods of estimation for confirmatory factor analysis with ordinal data. *Psychological Methods, 9*, 466–491.

Fox, J. (2008). *Applied Regression Analysis and Generalized Linear Models*. Thousand Oaks, CA: Sage.

Gorsuch, R.L. (1983). *Factor Analysis*. Hillsdale, NJ: Lawrence Erlbaum Associates.

Hosmer, D.W., Lemeshow, S., and May, S. (2008). *Applied Survival Analysis: Regression Modeling of Time to Event Data*. New York: John Wiley & Sons.

Hox, J. (2002). *Multilevel Analysis: Techniques and Applications*. Mahwah, NJ: Lawrence Erlbaum and Associates.

Hu, L. and Bentler, P.M. (1999). Cutoff criteria for fit indexes in covariance structure analysis: Conventional criteria versus new alternatives. *Structural Equation Modeling, 6*, 1–55.

IBM Corp. Released 2013. IBM SPSS Statistics for Windows, Version 22.0. Armonk, NY: IBM Corp.

Iversen, G. (1991). *Contextual Analysis*. Newbury Park, CA: Sage.

Kaplan, D. (2014). *Bayesian Statistics for the Social Sciences*. New York: The Guilford Press.

Kline, R.B. (2016). *Principles and Practice of Structural Equation Modeling*. 3rd ed., New York: The Guilford Press.

Kreft, I.G.G. and de Leeuw, J. (1998). *Introducing Multilevel Modeling*. Thousand Oaks, CA: Sage.

Kreft, I.G.G., de Leeuw, J., and Aiken, L. (1995). The effect of different forms of centering in hierarchical linear models. *Multivariate Behavioral Research, 30*, 1–22.

Liu, X. (2012). *Survival Analysis: Models and Applications*. Hoboken, NJ: Wiley.

McDonald, R.P. (1999). *Test Theory: A Unified Treatment*. Mahwah, NJ: Lawrence Erlbaum and Associates.

Mehta, P.D. and Neale, M.C. (2005). People are variables too: Multilevel structural equations modeling. *Psychological Methods, 10*, 259–284.

Muthén, B. (1989). Dichotomous factor analysis of symptom data. In Eaton & Bohrnstedt (Eds.), Latent Variable Models for Dichotomous Outcomes: Analysis of Data from the Epidemiological Catchment Area Program. A special issue of *Sociological Methods & Research, 18*, 19–65.

Pett, M.A., Lackey, N.R., and Sullivan, J.J. (2003). *Making Sense of Factor Analysis*. Thousand Oaks, CA: Sage.

Raudenbush, S.W. and Bryk, A.S. (2002). *Hierarchical Linear Models*. Newbury Park, CA: Sage.

Ryu, E. and West, S.G. (2009). Level-specific evaluation of model fit in multilevel structural equation modeling. *Structural Equation Modeling, 16*, 583–601.

Snijders, T. and Bosker, R. (1999). *Multilevel Analysis: An Introduction to Basic and Advanced Multilevel Modeling*. Thousand Oaks, CA: Sage.

Stapleton, L.M. (2013). Multilevel structural equation modeling with complex sample data. In G.R. Hancock and R.O. Mueller (Eds.), *Structural Equation Modeling: A Second Course* (pp. 521–562). Charlotte, NC: Information Age Publishing.

Therneau, T.M. and Grambsch, P.M. (2000). *Modeling Survival Data: Extending the Cox Model*. New York: Springer.

Wirth, R.J. and Edwards, M.C. (2007). Item factor analysis: current approaches and future directions. *Psychological Methods, 12*, 58–79.

Wooldridge, J. (2004). *Fixed Effects and Related Estimators for Correlated Random Coefficient and Treatment Effect Panel Data Models*. East Lansing, MI: Department of Economics, Michigan State University.

Index

Note: Page numbers followed by f and t refer to figures and tables, respectively.

Printed in the United States
by Baker & Taylor Publisher Services